Traditional Chinese Architecture Surveying and
Mapping Series:
Shrines and Temples Architecture

THE QUFU CONFUCIAN
TEMPLE

Compiled by School of Architecture, Tianjin University &
Qufu Municipal Administration of Cultural Heritage
Chief Edited by WANG Qiheng
Edited by WU Cong, DING Yao

China Architecture & Building Press

中国古建筑测绘大系·祠庙建筑

曲阜孔庙

国家出版基金项目
NATIONAL PUBLICATION FOUNDATION

『十二五』国家重点图书出版规划项目

天津大学建筑学院　曲阜市文物管理委员会　合作编写

王其亨　主编　吴葱　丁垚　编著

中国建筑工业出版社

Contents

目　录

Introduction: Shaping China for Thousands of Years

I

In 478 BC, Duke Ai of Lu State established a modest temple dedicated to Confucius who had passed away the year before based on the former residence of Confucius in Qufu. However, once the Temple was built, it has survived for almost 2,500 years so far, and became more and more magnificent with successive extensions. In comparison to some stone temple ruins of ancient Greece and Rome, the extreme longevity of the Temple of Confucius in Qufu is unexpectedly rare, which can be regarded as a symbol of the tenacious vitality of Chinese culture. Its long-lasting durability is not for its sound buildings, but due to the precious spiritual properties and heritage which Confucius left us. As the founder of Confucianism, a great thinker and philosopher, Confucius collected the great achievements of ancient Chinese culture, opened the door to hundreds of philosophers in the Warring States period (475 BC-221BC) and had occupied the top seat of ancient Chinese culture for more than two thousand years. This not only profoundly affected East Asian cultures, but also became one of the sources of thought for the Enlightenment in the West, which brought him the reputation of "shaping the world"; Confucius, as an educator and a Great Sage-Teacher, taught his disciples with in-discrimination, which broke the monopoly of knowledge and afterwards promoted the flow of social strata and social stability.

With the recognition of the authority of Confucianism, the changes of Confucius' posthumous titles, and the promotion of ceremonial specifications for Confucius, the Temple developed from its initial residence-based form, and underwent simple expansion of Han and Tang dynasties (206 BC-220AD, 618-907) and expansion of the Northern Song dynasty (960-1127) which shaped the layout of the Temple, and finally reached its peak in Ming and Qing dynasties (1368-1644, 1644-1911). The Temple today covers an area of around 9.5 hectares, about 652 meters long from north to south, with east-west width of 156 meters. The whole ensemble can be divided into six areas (fig. 1): 1) the three front courtyards ("the Great Court Wall" included), 2) the Star of Literature Pavilion and the courtyard in front of the Gate of Great Achievements, 3) the Main Court, 4) the East Court, 5) the West Court, and 6) the Backyard.

导言 经天纬地存千古

一

公元前 478 年，鲁哀公在阙里孔子故居因宅立庙，为前一年辞世的孔子建立了三间简陋的庙祀。然而，曲阜孔庙的生命历程一经开启便延绵未绝，且代有增益，渐趋恢宏，延续至今近 2 500 年仍活力充沛。对照已是华屋秋墟的古希腊、罗马的石作庙宇，曲阜孔庙的『寿与天齐』可谓世所罕见，诠释并象征着中华文化的顽强生命力，被当世列为『世界遗产』。曲阜孔庙之所以能永续不绝，非为其殿宇的固若磐石，而是源自孔子留给后世的宝贵精神财富和遗产。作为儒家学说创始人、伟大的思想家和哲学家，孔子集华夏上古文化之大成，开战国诸子百家之先河，并位居后世中华文化主轴地位两千余年，不仅深刻影响了东亚文化圈，而且也成为西方启蒙的思想资源之一，称『德牟天地』并非过誉；孔子作为教育家、『至圣先师』，有教无类，打破知识垄断，也促成后世的社会阶层流动和社会稳定。

随着儒学统治地位的确定、孔子封谥的变化和祭祀规格的提升，孔庙从最初以宅为庙，经历了汉唐简略，北宋扩建而奠定格局，至明清臻于极盛。现有的曲阜孔庙占地约 9.5 公顷，南北长约 652 米，东西宽约 156 米。南北狭长，院落重重（图一），整体可分为六个区：（一）前庭三院（附万仞宫墙）；（二）奎文阁和大成门前院；（三）正殿殿庭；（四）殿庭东院；（五）殿庭西院；（六）殿庭后院。

图一 曲阜孔庙鸟瞰图，从大中门到大成殿

Fig.1 Aerial view of Temple of Confucius, from the Gate of the Mean to the Hall of Great Achievements

The three front courtyards are spacious and solemn with sparse buildings on the central axis, including, from south to north, Jinsheng Yuzhen Archway, Lingxing Gate (Fig. 2), the Timeliness of the Sage Gate, Bishui River (Pan River) and Bishui Bridge, the Gate of Developing the Philosophy, the Gate of the Mean. In addition, outside Lingxing Gate, Gate of Respecting the Sage for Qufu City, usually regarded as the Temple's "Great Court Wall", can be seen as the starting point of the most southern end of the Temple.

The next area includes the Unity of Written Language Gate, the Star of Literature Pavilion which serves as a gateway as well as a library, and more than a dozen of Stele Pavilions successively built from Jin dynasty (1115-1234) to Ming and Qing dynasties.

Inside the Gate of Great Achievements, the buildings are divided into three groups: the Main Court, the East Court and the West Court. The Main Court is the core ceremonial area of the Temple, which includes the Gate of Great Achievements, the Apricot Platform, the Hall of Great Achievements (Fig. 3), the Resting Hall and the side corridors. It is believed that the East Court used to be the "old residence" of Confucius. This court includes the Gate of Sages, the Hall of Poetry and Rites, the Wall of Lu, the Well of the Old Residence, the Shrine of Sages, the Ancestral Temple and so forth. And the West Court is made up of the Gate of Sage Enlightenment, the Hall of Harmonious Music, the Hall of Sage Enlightenment King and its Resting Hall, which are dedicated to Confucius's father.

The main building of the Backyard is the Hall of Confucius's Deeds which exhibits the deeds and portraits of Confucius. There are also some auxiliary buildings such as the sacred kitchen, the sacred abattoir and the God of the Earth Shrine.

In addition to the ancient buildings, there are hundreds of tablets, plaques, inscriptions and other relics preserved in the Temple.

II

The basic function of the Temple was sacrificial worship. It could hold thousands people for large-scale rituals, with propitious, magnificent and sacred atmosphere around, and spacious courts and immense platforms. In history, Confucius got many posthumous names or titles such as Ni Fu, Uncrowned King, Extremely Sage Departed Teacher, King of Wenxuan and Emperor Wenxuan. Hence, the Temple has the "emperor's quality" which matches the status of his posthumous titles and the level of sacrifice, reflected in the "halberd-decorated gate" system, "five-gate system of the Son of Heaven," the system of front halls and rear residence, the configuration of corner towers of king cities, and even the use of the dragon pillars.

前庭三院建筑疏朗肃穆，由南而北包括金声玉振坊、棂星门（图二）、圣时门、璧水河（泮河）与璧水桥、弘道门、大中门等中轴线上重要建筑。另外，棂星门外的曲阜城仰圣门常被当作孔庙的『万仞宫墙』，可视作孔庙最南端起点。

奎文阁和大成门前院，包括同文门，作为藏书楼兼作殿门的奎文阁、金至明清时期逐渐密集加建的十多座碑亭以及东路斋宿院等建筑。

大成门以内，建筑组群分中、东、西三路，即殿庭、殿庭东院和西院。殿庭是孔庙核心祭祀区，包括大成门、杏坛、大成殿（图三）、寝殿及两庑，殿庭东院是孔子的『故宅区』，设有崇圣门、诗礼堂、鲁壁、故宅井、崇圣祠、家庙等；西院包括启圣门、金丝堂、启圣王殿、寝殿等，后两者用于祭祀孔子父母及五代祖。

殿庭后院中轴线为展示孔子事迹和画像的圣迹殿，并建有神厨、神庖、后土庙等附属建筑。

另外，整个孔庙之内，除古建筑外，还有数百碑碣、匾额、题刻等传世文物。

二

孔庙功能首在祭祀。佳气绕墙，庙貌岩岩，圣域幽奥，殿庭宽阔，月台宽敞，承载着多达千人的大规模祭祀仪典。孔子历史上尊称或封谥为『尼父』『素王』『至圣先师』『文宣王』『文宣帝』等，因而，孔庙自宋金以降逐渐形成的门载之制，『天子五门』、前殿后寝、王城角楼的配置，乃至龙柱的使用，使孔庙具有『帝王』气象，与孔子封谥曾居帝王之位、祭祀最高到『太牢』规制相应。

图三 大成殿和祭孔活动

图二 金声玉振坊和棂星门

Fig. 2　Jinsheng Yuzhen Archway and Lingxing Gate
Fig. 3　The Hall of Great Achievements and the memorial ceremony of Confucius

Walking in the Temple, visitors can appreciate the magnificent ceremonies, sacred sanctity, profound thoughts, the veneration to Confucius by people, and the majestic halls and various forms of buildings. At the same time, through the accumulation and enhancement of the Temple in the past dynasties, visitors could read the history of the stormy years, the admirable great sages, the stories of enlightenment and inspiration. This owes to some specially designed buildings and features in the Temple other than those which meet the needs of the memorial ceremonies and ritual symbols. These features can be seen as a part of a *guji* system of the hometown of Confucius, including the Great Court Wall, River Panshui Bridge, the Star of Literature Pavilion, the Apricot Platform (Fig. 4), the Hall of Harmonious Music, Hall of Poetry and Rites (Fig. 5), the Wall of Lu, the Well of Old Residence(Fig. 6), the Hall of Confucius's Deeds and even the juniper planted by Confucius (Fig. 7), enriching the space of the Temple, and interpreting and disseminating Confucianism. Among them, the Court Wall, River Pan Bridge and the Star of Literature Pavilion (also called the Pavilion of the Six Classics, the Library Pavilion, or the Pavilion of Classics, etc.) became a universal configuration of the Confucian temples throughout China. This is closely related to the unique conception of *guji* and the tradition of *guji* interpretation in ancient China.

Most people today would like to cherish and protect their cultural heritage, and similarly, ancient Chinese society had developed a tradition of appreciation of "*guji*." *Guji* literally means ancient traces or remains left to us by our ancestors. However, relevant research shows that *guji* contains the profound connotation of "human footprints" and "traces of age." It is not only the chanting of the literati while traveling, but also the systematization of value recognition and even systematic construction. *Guji* is the ancestors' activity or behavior traces, which, with the elapse of time and participation by generations of their decedents, has been accumulated and forged to become a kind of collective memory and composite cultural heritage. Early and clear records of *guji* can be found in *Book of Han* (ca. 3rd century). Since Song dynasty (960-1297), and most *guji* sites were recorded collectively and systematically in local records (*fangzhi*) as an independent section. Since the 13th century, at least, relatively complete *guji* inventories were kept and continuously updated almost in every county throughout China. Some were even presented in town/village level of local records, and at the same time, there were a lot of special *guji* records for landscape areas, Buddhist temples as well. Intrinsically, the core attributes of *guji* is humanism, and the soul of *guji* lies in humans. *Guji* is supposed to be linked with human's deeds, morality, feelings, and interpersonal relationships. *Guji* sites and their related historical texts are complementary to form an integral part; they are the coordinates in the historical framework of space and time; and *guji* is a blend and crystal of human historical development with their geographical background.

走读孔庙，固然可以看到礼制之威仪、圣域之肃穆、思想之深邃、后世之尊崇、殿宇之宏敞，启蒙励志的故事，等等。这有赖于曲阜孔庙中专门兴修了一些特有的建筑和景物，在满足祭祀和礼仪象征的建筑之外，构成了孔子故里『古迹』体系的一部分，包括万仞宫墙、泮水桥、奎文阁、杏坛（图四）、金丝堂、诗礼堂（图五）、鲁壁、孔宅故井（图六）、圣迹殿，乃至孔子手植桧（图七）和杏树等，从形式上丰富了孔庙的空间，内涵上阐释并传播了儒家思想。其中，宫墙、泮水桥和奎文阁（或称六经阁、御书阁、尊经阁等）成为各地文庙通行的配置。这与中国古代特有的古迹观念和擅长阐释古迹的传统密切相关。

然而历代孔庙的踵事增华和深厚积淀，则更让人读出风雨经年的历史，高山仰止的圣贤、殿宇之宏敞，建筑之丰富，人悠游时的吟咏，更是体系化的价值认知甚至制度建设。古迹是先人活动和行为的遗存，是经过时间的洗礼，在后世人们的参与下，叠加、积淀而形成的集体记忆和复合文化遗产。早期比较明确的古迹记录见于《汉书·地理志》，宋代以降，古迹则大多在方志中集中、系统记录，并成为独立的门目。至迟到宋元以降，全国在县一级行政单位内，基本上都有一份比较完备的、由官方认定并持续更新的古迹名录。有些村镇一级方志中也出现了古迹名录，同时还伴随着大量的古迹专志的出版。地以人胜，事以文传。从内涵上说，古迹的核心属性是人，古迹的灵魂是人，系人事、赞人品、感人情、思事理。古迹与史志相表里，是历史发展时空框架中的坐标，是自然地理背景与人文历史发展的交融和绝响。

与当今世界各国均珍视和保护的『文化遗产』类似，中国古代社会一直有珍重『古迹』的传统。古迹，字面上看是先人遗迹的统称。但相关研究表明，古迹更包含了『人类脚印』『岁月痕迹』的深刻内涵，不仅是文人悠游时的吟咏，

Fig. 4　The Apricot Platform
Fig. 5　The Hall of Poetry and Rites

图七　孔子手植桧

图六　孔宅故井和鲁壁

Fig. 6　The Well of Old Residence and the Wall of Lu
Fig. 7　The juniper planted by Confucius

It seems that in the past focus was not on the "preservation" of *guji* sites, but on the interpretation and dissemination of them historically, ethically and artistically. In Ming dynasty, Qufu was a county of Yanzhou Prefecture. In *the Local Records of Yanzhou Prefecture* (compiled in Wanli period), the introduction to the *guji* section discusses with pride the values of *guji* and its interpretation:

Guji and the classical works of Confucius are complementary. It can be interpreted gloriously, splendidly and vividly via different media and in different forms of art, such as words of thinkers, inscriptions, folk songs, prose and poetry, publications and historical archives

This statement on *guji* represents not only the historical and social values of *guji*, but also the general attitudes and practices about *guji* in ancient China: 1) with its status as high as that of Confucius' works, *guji* was regarded as a classic, precious object, and it is complementary with historical texts, 2) ancient Chinese people specialized in "interpretation" of *guji* with literature works, inscriptions, prose and poetry, paintings and other media, which rendered *guji* sites into beautiful attractions which are deep in people's mind, resulting in the universal dissemination and inheritance of *guji* values for later generations.

Related Qufu local records, archives of Kong family and a large number of related inscriptions, prose and poems show that, in people's mind, *guji* sites are not isolated in a scattered manner, but constitute a hierarchic system, from the distant to the near, from the large to the small, integrating into the time-space framework of the local historical evolution, and the origins and development of the "sacred deeds" of Confucius and Confucianism. Specifically, in the hometown of Confucius, on the macro level there are the East See and Mount Tai, Zou and Lu, the mountains of Tai and Dai, Gui and Meng, Fu and Yi, Ni and Fang, and the Rivers of Zhu, Si, Wen and Yi; on the medium level, there are the Temple and Tomb of Confucius, the Site of the Lu State Capital, Queli (the neighborhood of Kong family) and the Mean Alley (where Yan Hui was supposed to live); and on the micro level, there are Lingguang Hall, the Apricot Platform, the Star of Literature Pavilion, the Hall of Poetry and Rites, and even the Wall of Lu, the Well of Old Residence, the juniper planted by Confucius and Confucius' clothing and sso on.

The so-called "*guji*" within Temple of Confucius can be divided into two categories according to their links with the dates before or after Confucius' death: 1) the symbol of the "sacred deeds" of Confucius, e.g. the juniper planted by Confucius, the Apricot Platform and the apricot trees, and the Hall of Poetry and Rites; 2) the symbol of Confucianism's indomitable vitality, e.g. the Well of the Old Residence, the Wall of Lu and the Hall of Harmonious Music, which were used to interpret the story that books were burned out and alive Confucian scholars were buried during Qin dynasty, while

诚然，传统上对于古迹并未聚焦在实物的『保存』，而是重视对古迹的『阐释』和传播，从历史发展、社会伦理以及艺术创造等方面用不同方式进行阐释。明代曲阜县隶属兖州府，关于对古迹的价值以及对古迹的阐释、传播，在（万历）《兖州府志》关于古迹的序言中自豪地写道：

左史纪言，迹载六经；右史纪动，迹具春秋……六经之遗，春秋之续，古人言动之迹，散百家、勒晶彝、镌金石、歌民风、咏诗赋、书简牍、垂竹帛而藏名山者，缤纷乎珠玉并陈，彪炳乎丹青焕烂。

这段古人对待古迹的代表性论述，不仅指出古人心中古迹的历史和社会价值，更折射出古人对待古迹的态度和做法：一是将古迹视为经典，珍如珠玉，与史志互补，地位之高犹如孔子修订的六经；二是擅长对古迹进行『阐释』，用百家著作、金石碑铭、诗词歌赋、丹青彩绘等各种媒介载体，将古迹演绎表现得绚烂多姿，深入人心，以追求古迹价值的普世传播和世代传承。

研读曲阜方志、阙里志和大量相关碑铭诗文，可以看出古人心目中的各处古迹并非散乱分布的孤立存在，而是构成了由远及近，从大到小，层次丰富的体系，融于历史进程的时空框架与孔子『圣迹』和儒家思想的源流演变。具体到孔子故里而言，宏观层次则有海岳、邹鲁，山有泰岱、龟蒙、尼防，水有洙、泗、汶、沂；中观层次则有孔子林庙、鲁国故城、阙里、陋巷；微观层次则有鲁灵光殿、杏坛、奎文阁、诗礼堂，乃至鲁壁、故井、手植桧和孔子衣冠等。

位于孔庙之内的所谓『古迹』，按孔子生前和身后可以分为两类：一是反映孔子『圣迹』的孔子手植桧、杏坛、杏树和诗礼堂；二是借孔宅故址，用以阐释诗书焚燕、鲁壁藏书，恭王坏宅等曲折故事的故井、鲁壁和金丝堂等，反映儒家经典历乱世而得以保存并重见天日，从而象征孔子思想顽强生命力。从内涵上看，这些古迹又可分为

some copies of Confucian classics survived the disasters and were rediscovered in Han dynasty, owing to Confucius' descendants who hid and preserved the books in a wall of their house. According to the connotation, these *guji* sites can also be divided into several categories, such as the Hall of Poetry and Rites, the Wall of Lu and Hall of Harmonious Music, reflecting the great philosopher and thinker Confucius; the Apricot Platform reflecting the educator Confucius, and the juniper which was believed to be planted by Confucius and retain Confucius' luster. Among them, except for the juniper and the well which remains seem to be traceable more or less, most futures were obviously created by later generations, hoping to interpret these *guji* sites in a representative and symbolic way. The interpretation mode in Temple of Confucius began in Northern Song dynasty, when, with the expansion of the Temple, the Apricot Platform were built in the central axis, and the Hall of Harmonious Music was added in the East Court. In Hongzhi period of Ming dynasty, the Hall of Harmonious Music was moved to the West Court, while in the East Court, the composition of the Hall of Poetry and Rites, the Wall and the Well was formed. In addition, there are more than 100 pieces of stone carvings of picture-story of Confucius's deeds, Confucius' portraits, and the maps of Confucian ceremonies by Emperor Gaozu of Han and Emperor Zhenzong of Song, preserved in the Hall of Confucius's Deeds which was built in Wanli period of Ming dynasty. This is the interpretation and presentation of the Confucius and related stories with images.

With the interpretation of these sites, a unique *guji* system was gradually developed from Northern Song dynasty to Ming and Qing dynasties in the Temple, a place with worship functions, ceremonial spaces and commemorative symbols: it possesses the vividness of specific situations and creates a series of unique artistic images and atmosphere; It goes directly into people's mind arousing meditations, as a Chinese saying explains: "the traces of the ancients, the heart of the ancients." It serves as key nodes to reconstitute the macro-historical context of Confucianism, enhancing the monumental and ceremonial natures of the Temple. Looking back on history, we are surprised to find that the interpretation of heritage turned out to be what we used to specialize in, an excellent tradition that we lose today.

Needless to say, these *guji* sites can not be used as historical testimony for the "sacred deeds" of Confucius. However, history is not one moment, but every moment, as Emperor Qianlong asked in his poem about the Wall of Lu, "can't the moment when the books were rediscovered stand for (Confucius' great achievements) of shaping the world for thousands of years? " History is a process of development, generations after generations. "Our descendants will look on us as we look on our ancestors." Therefore, these *guji* sites are still the historical testimonies of the ancients' prestige of Confucius and spreading of Confucianism, telling a story that ancient Chinese people specialized in interpreting and disseminating the ideology and culture with dynamic conception of *guji*, as well as giving us inspiration for the interpretation and presentation of the heritage today.

若干类别：如体现思想家、哲学家孔子的诗礼堂、鲁壁、金丝堂，反映教育家孔子的杏坛，余留孔子手泽的『手植桧』等。其中除手植桧、故井遗址或略有迹可循，其余景物显系后世创建，是一种采用情景再现和象征手法的古迹『阐释』方式。孔庙的阐释模式始于北宋，随着孔庙规模的扩大，中轴线上兴建杏坛，东路兴建金丝堂等，并在明弘治时期的扩建中，金丝堂移至西路，东路则形成了诗礼堂、鲁壁、故宅井等格局。另外，在始建于明万历年间的圣迹殿内，还保存着百余幅石刻圣迹图、孔子圣像、汉高祖和宋真宗祭庙图等，则是用图像方式对孔子相关历史的阐释和演绎。

利用对这些古迹的阐释，北宋至明清的曲阜孔庙，逐渐于祭祀功能、礼仪空间、纪念象征之外建成了一套特有的古迹体系：它具有具体情境的形象性和生动性，营造了特有的意象或意境；它直指人心，发人深省，所谓『古人之迹，古人之心也』；它作为关键节点构筑了孔子及其思想的宏观历史语境或称历史文脉，烘托了孔庙的纪念性和礼仪性。回首历史蓦然发现，遗产的阐释原来是我们曾经的优势和专长，是我们迷失方向而丢失了的优秀传统。

诚然，在今天看来，这些『古迹』显然无法作为孔子『圣迹』真实的历史见证。但是，正如乾隆皇帝在鲁壁题诗中的反问，『经天纬地存千古，岂系恭王坏宅时？』历史不是一时一刻，而是每时每刻。历史是一个选代发展的过程，『今之视昔犹后之视今』，因此，这些古迹仍然是古人对孔子崇誉有加、对儒家学说发扬传播的历史见证，是古人动态发展的古迹观念和善于阐释古迹、传承思想文化的生动写照，也为我们今天开展遗产的『阐释与展示』提供了启发。

III

From 2001 to 2004, the Qufu Municipal Commission of Cultural Heritage (the predecessor of the Qufu Municipal Administration of Cultural Heritage, QMACH) cooperated with the School of Architecture, Tianjin University to conduct a comprehensive recording project on the Temple of Confucius, Kong Family Mansion, the Cemetery of Confucius and His Descendants, the Temple of Zhou Gong, the Temple of Yan Hui, the Tomb of Shaohao, Zhu-Si Academy, the Confucian Temple in Mount Ni and other cultural heritage sites, in which measured survey of the Temple of Confucius was the first sub-project. In 2001, taking advantage of a teaching program of Tianjin University, "Field Trip for Measured Survey of Ancient Chinese Buildings" (later rated as the National Quality Courses), and under the leadership of Professor Wang Qiheng, some teachers and graduate students formed a teaching and guide group to organize the massive survey operation. All the undergraduate students of Grade 1999, majoring in architecture or city planning acted as the main force, as well as some students of Grade 1996 and 1998. Some undergraduates of Grade 2000 participated in the next year's supplementary surveying. There were totally about 120 teachers and students who participated in the survey work (see the name list of participants for details). Many teaching plans, cases, measured drawings and digital technology experience accumulated during the work were soon included in the textbook *Measured Survey of Ancient Buildings* edited by Professor Wang Qiheng and published in 2006. This textbook has filled the blank of the systematic technical and practical guides for measured survey of traditional wood buildings in East Asia in the past hundred years. It has been widely populated among the domestic heritage conservation institutions and professionals and was recommended as a training material for examination of responsible engineers for the heritage conservation in China.

The project of measured survey of ancient building in Qufu was co-initiated by Prof. Wang and Mr. Xu Ke, the then deputy mayor of Qufu and the then CPC Secretary of QMACH and Mr. Chen Chuanping, the then director of the QMACH. It was also supported by other vice directors of QMACH at that time: Ding Chen, Cheng Shaomin, Xu Huichen, Zhao Jiqun and Xiang Chunsheng. Mr. Kong Xiangfeng, the then administrative office director of QMACH was in charge of the specific implementation of the project, and other officials and staff members also gave great support and contributed to the project, such as Zhou Peng, Zhang Long, Kong Xiangmin, and Li Yuchun. The publication of this book also received the great attention and support by the current director of QMACH, Mr. Kong Deping, and QMACH contributed some additional photographic materials.

三

从 2001 年至 2004 年，曲阜市文物管理委员会（文物局前身）与天津大学建筑学院通力合作，开展了对曲阜孔庙、孔府、周公庙、颜庙、少昊陵、洙泗书院、尼山孔庙、尼山书院等文物保护单位的古建筑进行测绘记录的项目，曲阜孔庙古建筑的测绘是其中的首个项目。2001 年，结合天津大学『中国古建筑测绘』教学实践（后被评为国家级精品课程），在王其亨教授的领导下，由多名教师和研究生组成教学指导组，以 1999 级建筑学、城市规划全体本科生为主力，吸收部分 1996 级、1998 级本科生开展了声势浩大的曲阜孔庙的测绘工作。部分 2000 级本科生参与了次年的补测工作。先后参与测绘工作的师生约 120 人（详见测绘人员名单）。在曲阜古建筑的测绘过程中所积累的很多教案、测绘案例、测绘成果以及数字化技术经验等，收进了由王其亨教授主编、2006 年出版的国家普通高等教育『十五』规划教材《古建筑测绘》。这本教材填补了百余年来东亚传统木结构建筑测绘技术和操作规范的空白，在国内建筑院校和文化遗产保护专业人员中得到普遍应用，成为国家文物保护工程责任工程师考试指定教材。

曲阜古建筑测绘项目，是在时任曲阜市副市长、曲阜市文物局党委书记许可先生，以及时任文物局长陈传平先生邀请和支持下开始实施的，并得到文物局副局长丁晨、程少珉、徐会臣、赵吉群和项春生先生大力支持，提供了有利条件。具体负责组织实施测绘相关工作的是时任文物局办公室主任孔祥峰先生；另外，时任孔庙管理处主任周鹏，文物局办公室张龙、古建队孔祥民，文物科李玉春以及很多相关工作人员，都对测绘给予了大力支持，保障了测绘项目顺利实施。本书的出版，也得到现任曲阜市文物局长孔德平先生高度重视和支持，文物局方面又另外提供了部分摄影资料。

IV

Too much repairs and less research and interpretation turns out to be the weak point of the conservation of cultural heritage in China. It should be noted that the conservation of architectural heritage is not just "repairing" a building. In *Principles for the Conservation of Heritage Sites in China*, conservation refers to "all activities carried out to preserve the physical remains of sites, their historic settings and other relevant elements." The conservation procedures require that the value assessment should be taken as top priority, and research should be done throughout the processes of conservation. It should be noted that "repair" as a technical measure is a mere means, and it is the ultimate goal and the "original intention" of conservation that the culture should be passed down and the values be deeply rooted in people's mind. Therefore, the conservation of cultural heritage is "to preserve the tangible and hand down the intangible," requiring more research, interpretation and mass communication. It must not cease at repair and should not simply extend the life expectancy of physical remains. The basis and the starting point of research is to recording the historic buildings, including the measured survey.

It should be noted that documentation is closely related to the "authenticity" of heritage. No documentation, no authenticity. Relevant research shows that the authenticity includes the tangible side as well as the intangible side which is described as the "credibility of information sources" in Nara Document on Authenticity. The two sides are interdependent and mutually complementary. The so-called "credibility of information sources" are mostly in the form of documentation. From this point of view, the measured survey and recording of architectural heritage is by no means limited to a local, "preliminary work" of a conservation project, but a global work. It is not a "disposable" work but runs through the entire process of conservation. Documentation is crucial to "preventive conservation" and "critical restoration" as well. With the full development of digitization and information technology, the previous professional documentation is only one step away from dissemination. Therefore, documentation is also an important pillar for the interpretation, presentation and dissemination of architectural heritage.

Thus, we would not like to make the digital measured survey achievements of the Temple of Confucius in Qufu become "one-offs" and are willing to take advantage of mass media to share them as the important media for readers to study and appreciate the Temple. We would like to use digitization and information technology to continue to make further research on the measured survey achievements, and in the future with informatization survey technology to represent the grandeur of the Temple in a brand new way.

四

重修缮而轻研究、轻阐释，正是当前国内建筑遗产保护的软肋。应当看到，建筑遗产保护并不等同于「修房子。《中国文物古迹保护准则》规定，保护是指「为保存文物古迹及其环境和其他相关要素进行的全部活动」，保护程序要求将价值评估置于首要地位，研究贯穿于保护行为始终。应该看到，「修」作为保护的技术措施，只是手段，而传承文化、让优秀的价值观深入人心才是保护的终极目的和「初心」。因此，文化遗产保护是「存迹传道」，需要更多的研究、阐释和大众传播，而不能止于修缮，不能止于简单延续物质遗存的寿命。而研究的基础和起点，则是为古建筑进行科学记录建档，包括测绘。

应当看到，测绘记录与遗产「真实性」息息相关，没有记录就没有真实性。相关研究表明，真实性包括有形的「真品」和无形的「真传」，即《奈良文件》所称「信息来源的确凿可靠」。两者互为表里，相互依存。而所谓真传即信息来源，主要以记录档案的形式存在的。从这个角度看，建筑遗产的测绘与记录，绝不仅仅是某个保护工程项目的「前期工作」，而是遗产保护的全局性工作；不是一次性的工作，而是贯穿保护全过程；记录对「预防性保护」和「研究性保护」至关重要。在数字化、信息化技术充分发展的条件下，原来专业的记录工作与普及性的「传播」也只有一步之遥，因此记录也是建筑遗产阐释、展示和传播的重要支柱。

因此，我们不愿意让数字化的曲阜孔庙的测绘成果成为「一次性用品」，愿意利用大众传媒分享出来，成为广大读者研究、欣赏曲阜孔庙建筑的重要媒介，并愿意利用数字化、信息化技术，继续深化测绘研究成果，将来以信息化测绘的全新表达方式再现孔庙的辉煌。

The beginning of the construction: In 478 BC

Site area: 9.5 hectares

Competent organization: Qufu Municipal Administration of Cultural Heritage

Surveying organization: School of Architecture, Tianjin University

Time of surveying and mapping: 2001-2002

始建年代：公元前 478 年

占地面积：9.5 公顷

主管单位：曲阜市文物管理委员会

测绘单位：天津大学建筑学院

测绘时间：2001—2002 年

图

版

Drawings

中国古建筑测绘大系·祠庙建筑——曲阜孔庙

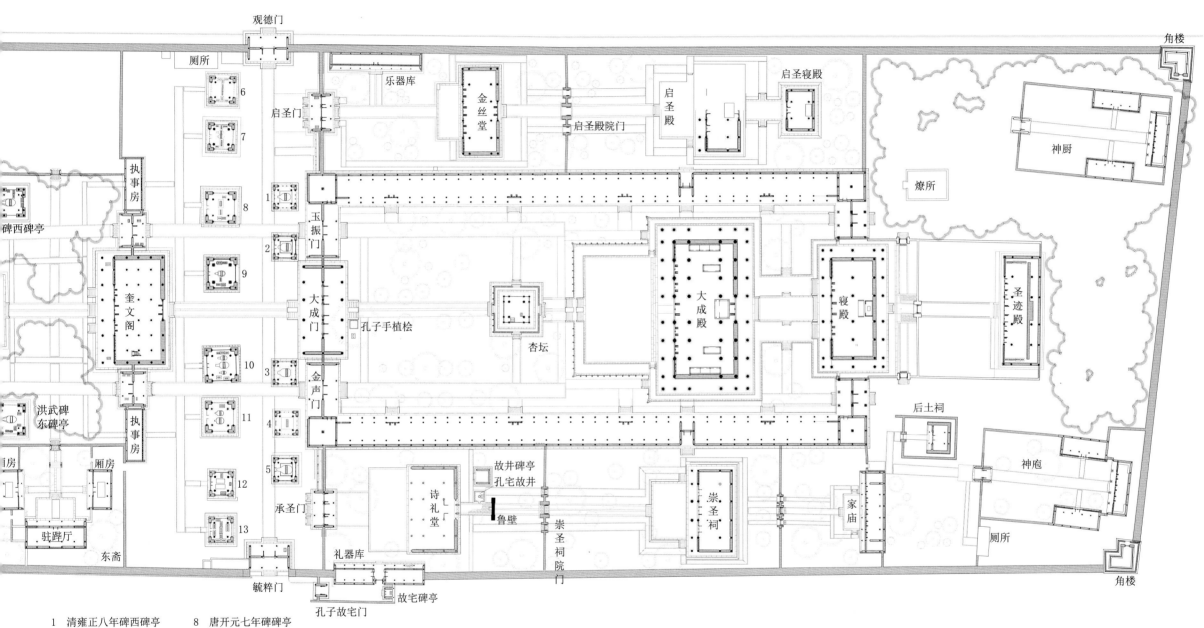

观德门

厕所

乐器库

金丝堂

启圣寝殿

启圣殿

启圣门

启圣殿院门

神厨

6

7

执事房

碑西碑亭

1

玉振门

2

燎所

9

奎文阁

大成门

孔子手植桧

杏坛

大成殿

寝殿

圣迹殿

3

金声门

10

洪武碑东碑亭

执事房

11

4

后土祠

5

诗礼堂

故井碑亭
孔宅故井

崇圣祠

家庙

神庖

厢房

鲁壁

12

承圣门

崇圣祠院门

东斋

13

驻跸厅

毓粹门

礼器库

厕所

孔子故宅门

故宅碑亭

角楼

角楼

1　清雍正八年碑西碑亭　　8　唐开元七年碑碑亭
2　清康熙三十二年碑碑亭　9　元至顺二年碑碑亭
3　清康熙二十五年碑碑亭　10　元至元五年碑碑亭
4　清雍正八年碑东碑亭　　11　宋太平兴国八年碑碑亭
5　清乾隆十三年碑碑亭　　12　清雍正元年碑西碑亭
6　清康熙十五年碑西碑亭　13　清雍正元年碑东碑亭
7　清顺治八年碑东碑亭

北

0　5　10　　20m

孔庙总平面图

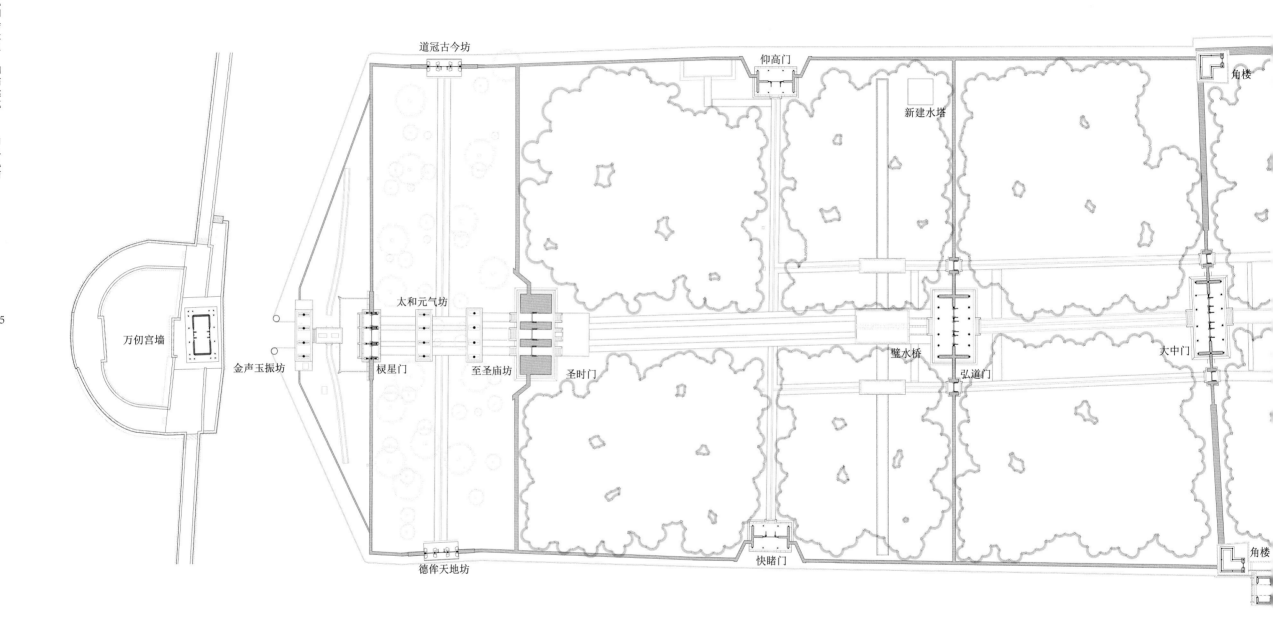

万仞宫墙

金声玉振坊

道冠古今坊

太和元气坊

棂星门

至圣庙坊

圣时门

德侔天地坊

快睹门

仰高门

新建水塔

璧水桥

弘道门

角楼

大中门

角楼

孔庙总平面图

Site plan of Confucian Temple in Qifu

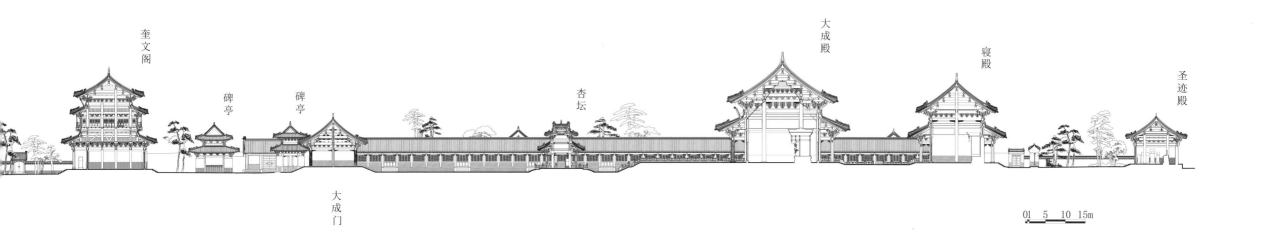

奎文阁

碑亭

碑亭

大成门

杏坛

大成殿

寝殿

圣迹殿

01 5 10 15m

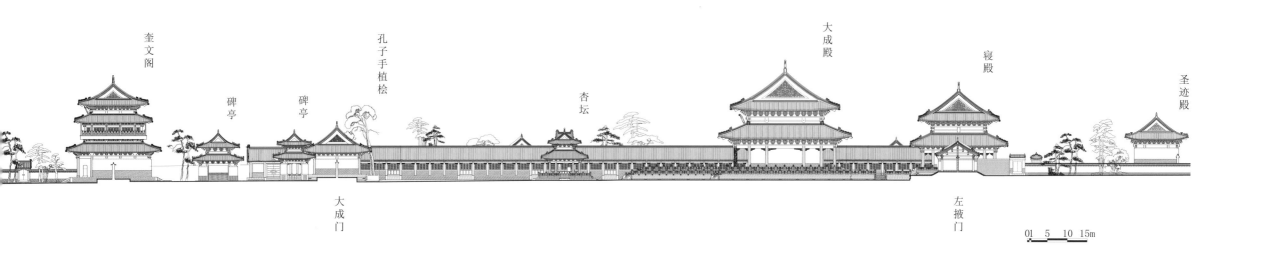

奎文阁

碑亭

碑亭

孔子手植桧

大成门

杏坛

大成殿

寝殿

左掖门

圣迹殿

01 5 10 15m

金声玉振坊　棂星门　道冠古今坊　圣时门　仰高门　弘道门　角楼

太和元气坊　至圣庙坊　璧水桥　大中门

孔庙组群纵剖面图
Longitudinal section of Confucian Temple building complex

金声玉振坊　棂星门　道冠古今坊　圣时门　仰高门　弘道门　角楼

太和元气坊　至圣庙坊　璧水桥　大中门

孔庙组群东立面图
East elevation of Confucian Temple building complex

前庭三院（附万仞宫墙）

The Three Front
Courtyards
(Enclose: Great
Court Wall)

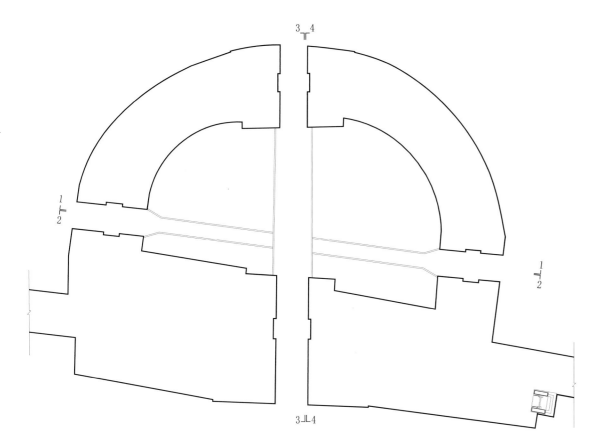

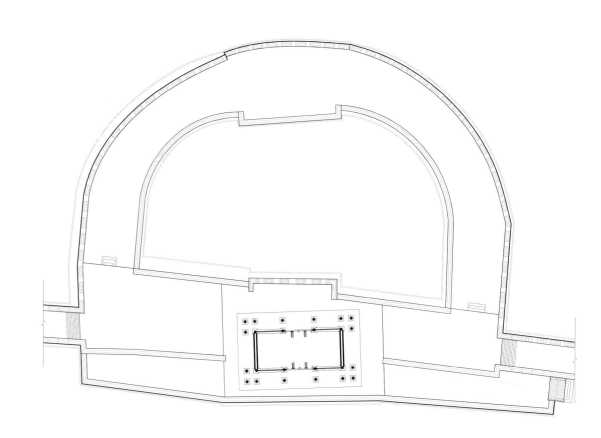

北

万仞宫墙一层平面图
Plan of first floor of Great Court Wall

万仞宫墙二层平面图
Plan of second floor of Great Court Wall

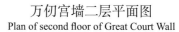
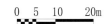

0　5　10　　20m

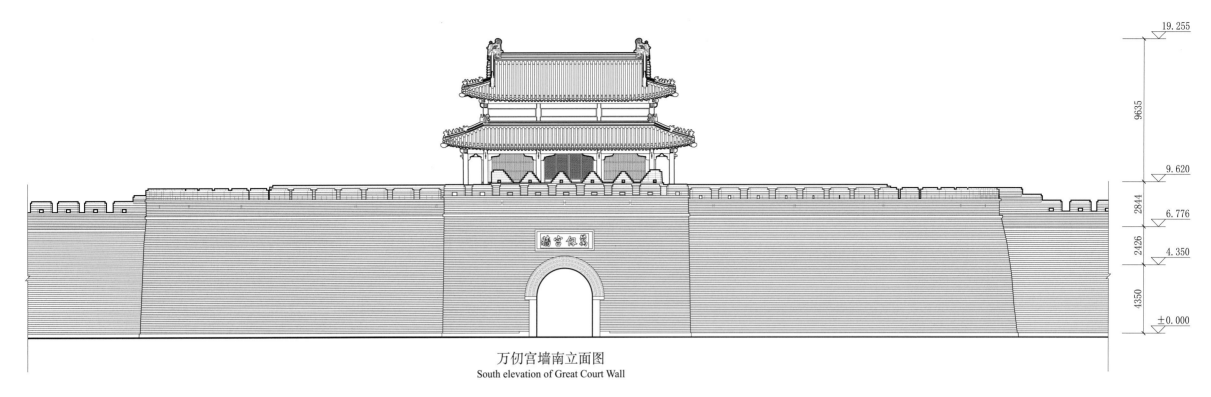

万仞宫墙南立面图
South elevation of Great Court Wall

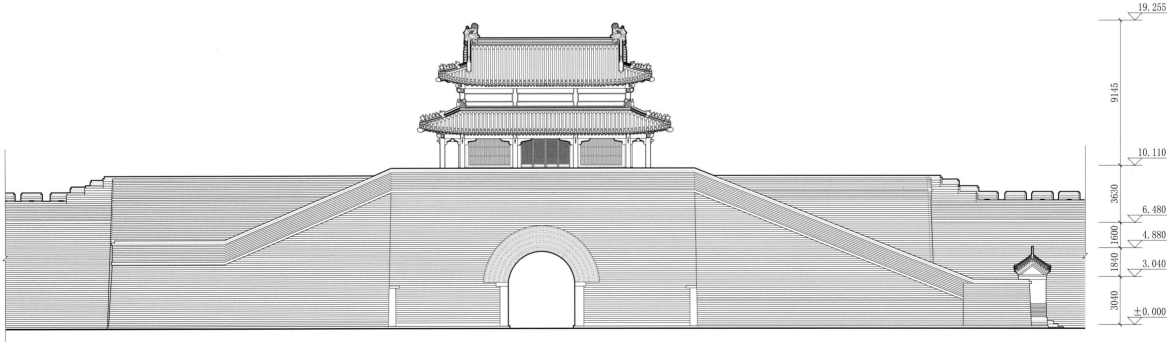

万仞宫墙北立面图
North elevation of Great Court Wall

0 1 5 10m

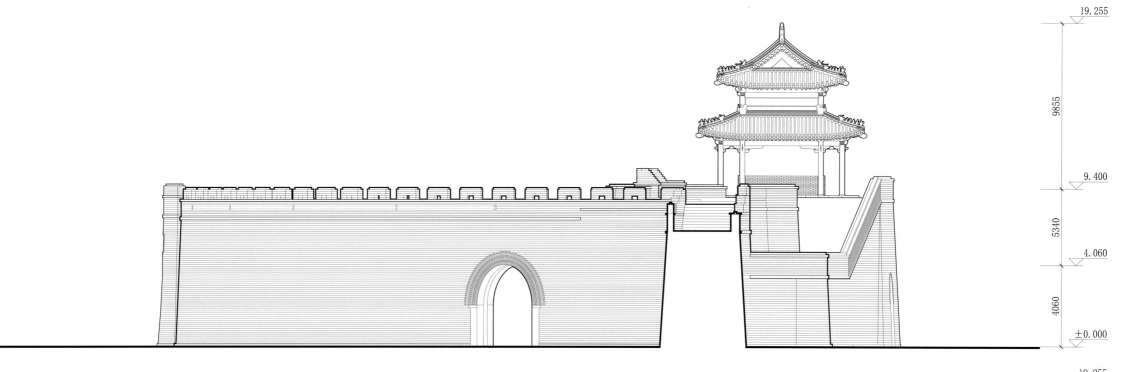

19.255

9855

9.400

5340

4.060

4060

±0.000

万仞宫墙东立面图
East elevation of Great Court Wall

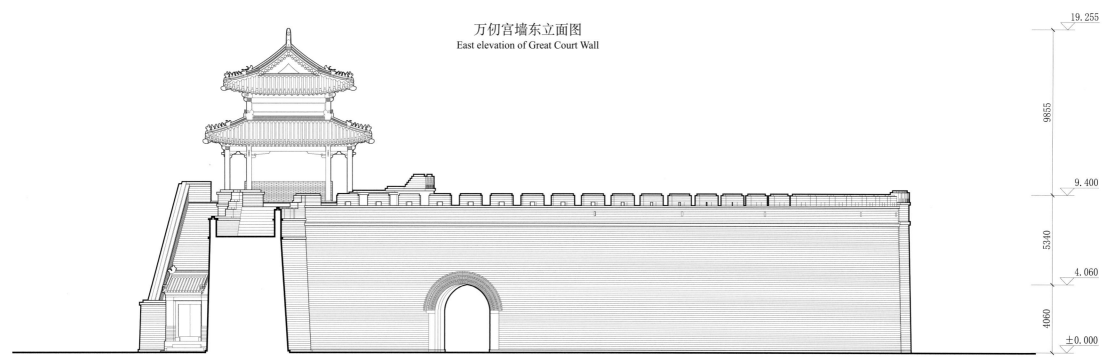

19.255

9855

9.400

5340

4.060

4060

±0.000

万仞宫墙西立面图
West elevation of Great Court Wall

0 1 5 10m

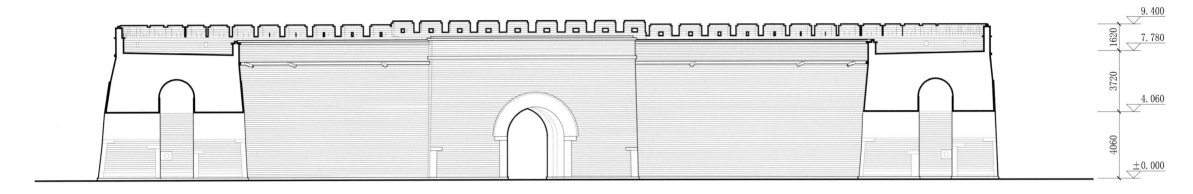

	9.400
1620	7.780
3720	4.060
4060	±0.000

万仞宫墙 1-1 剖面图
Section 1-1 of Great Court Wall

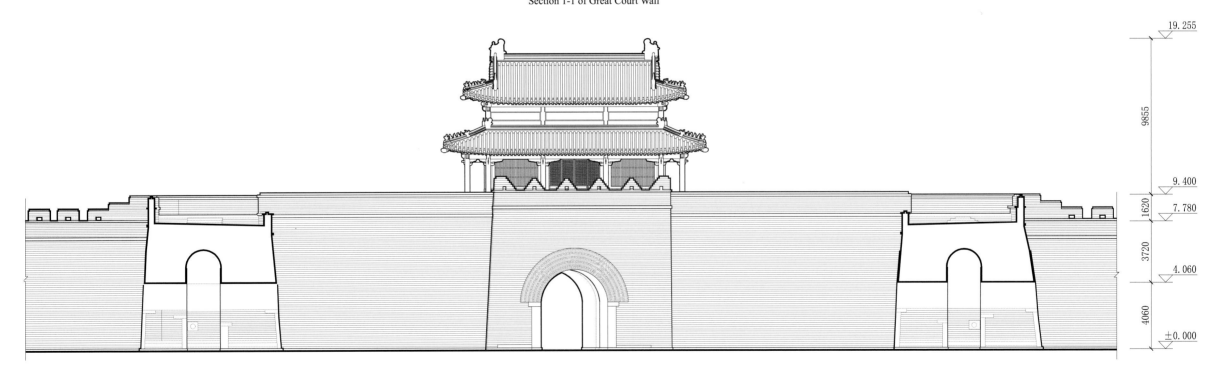

	19.255
9855	
	9.400
1620	7.780
3720	4.060
4060	±0.000

万仞宫墙 2-2 剖面图
Section 2-2 of Great Court Wall

0 1   5   10m

19.255

9145

10.110
1265 8.845

3965 4.880

4880

±0.000

万仞宫墙 3-3 剖面图
Section 3-3 of Great Court Wall

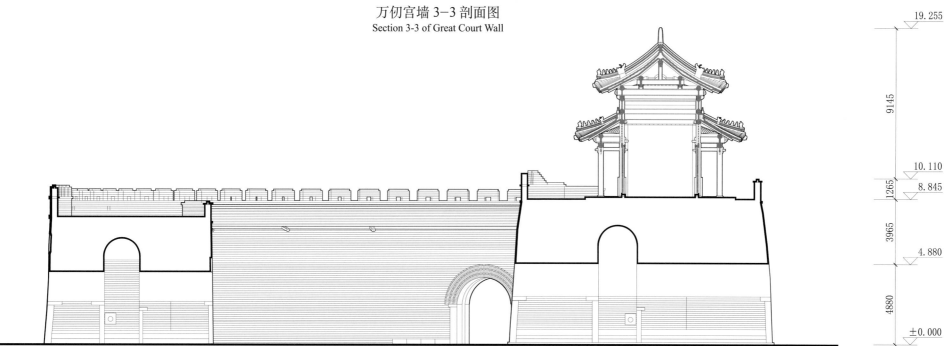

19.255

9145

10.110
1265 8.845

3965 4.880

4880

±0.000

万仞宫墙 4-4 剖面图
Section 4-4 of Great Court Wall

0 1   5   10m

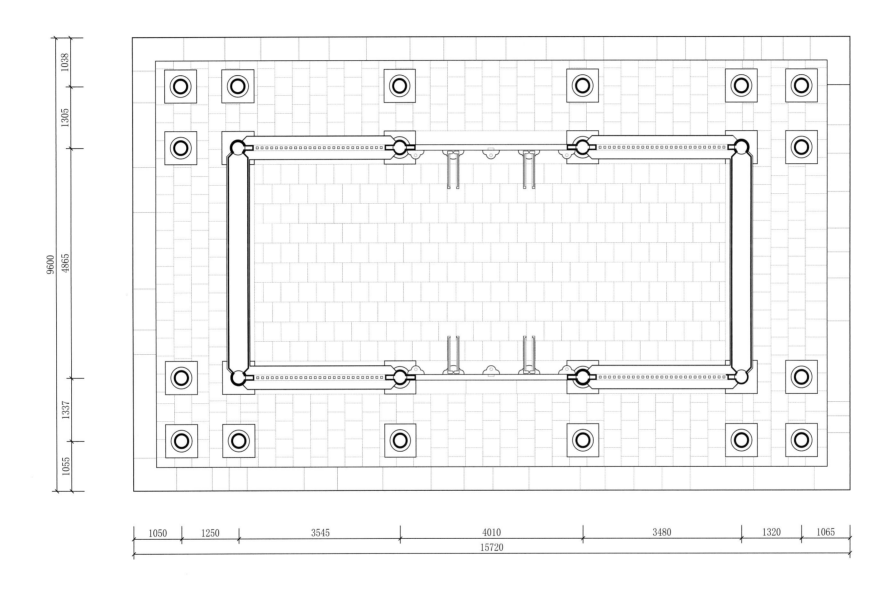

北

万仞宫墙城楼平面图
Plan of gate tower of Great Court Wall

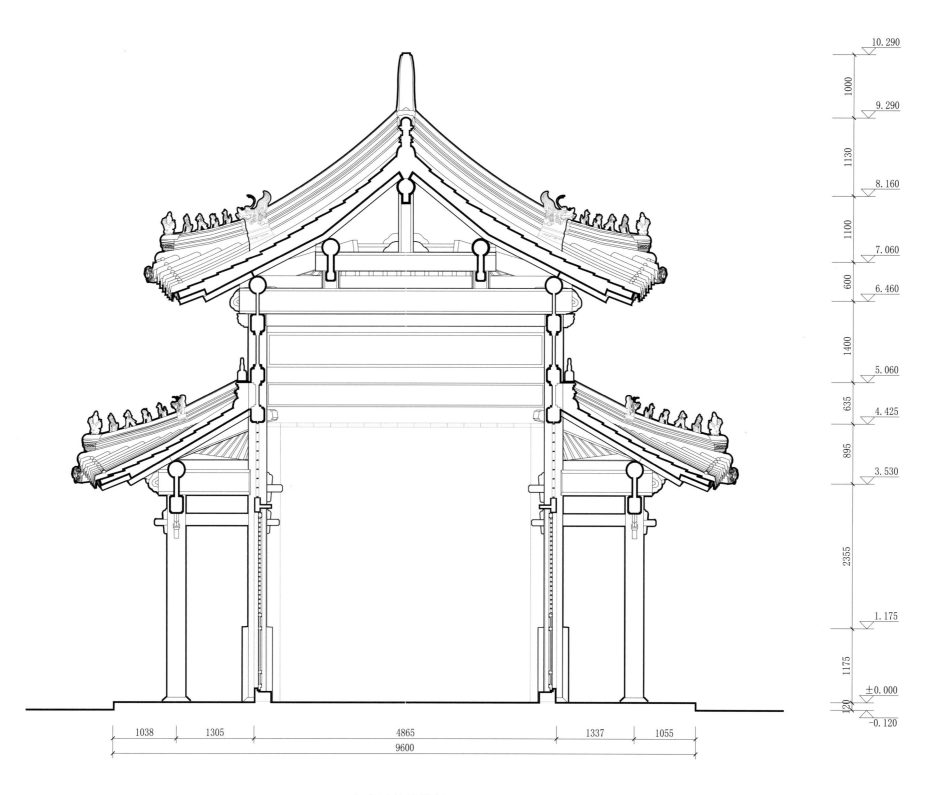

万仞宫墙城楼横剖面图
Cross-section of gate tower of Great Court Wall

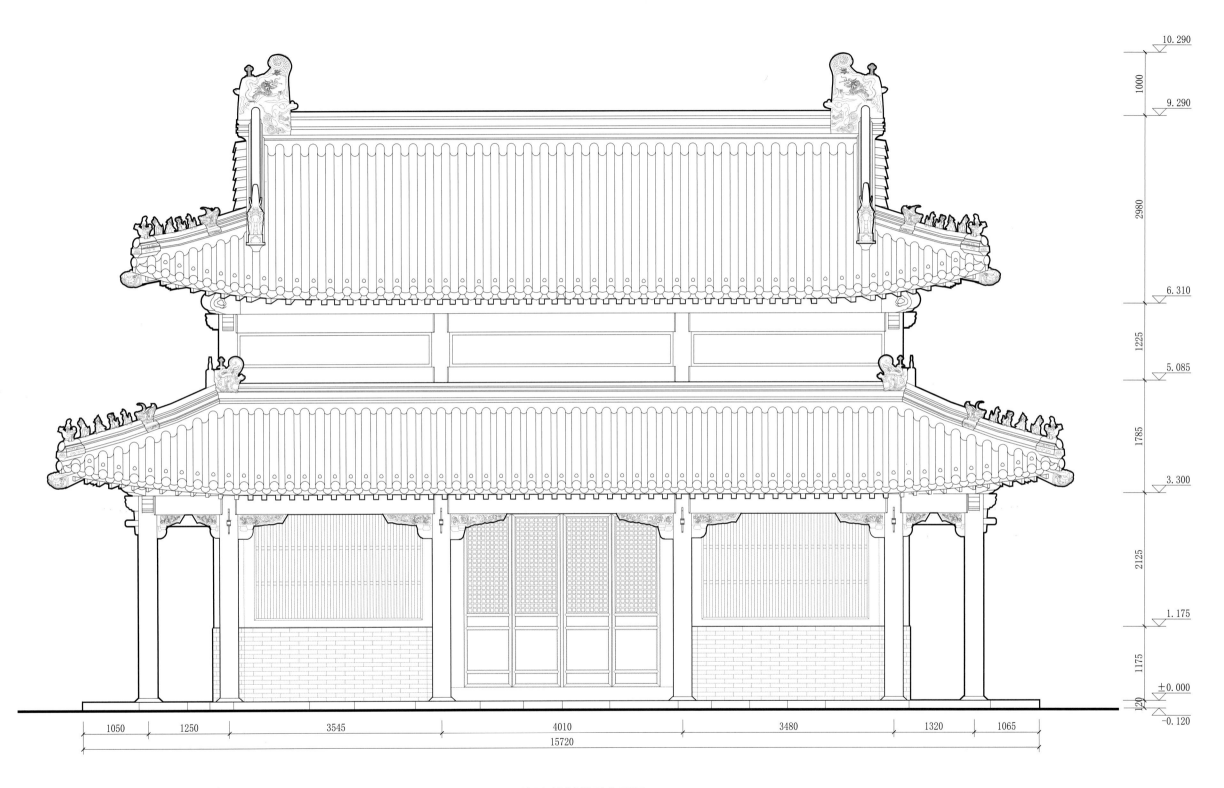

10. 290

1000

9. 290

2980

6. 310

1225

5. 085

1785

3. 300

2125

1. 175

1175

±0.000

120

-0.120

1050　1250　3545　4010　3480　1320　1065

15720

万仞宫墙城楼正立面图
Front elevation of gate tower of Great Court Wall

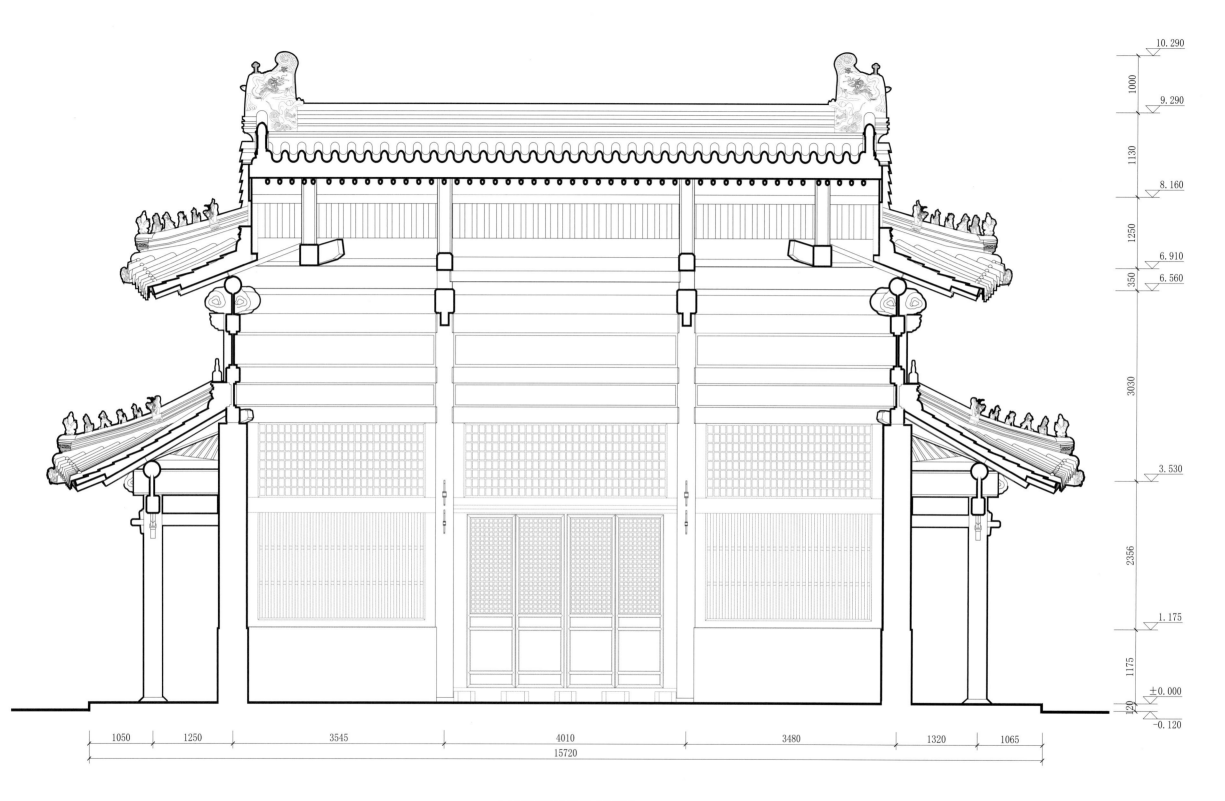

万仞宫墙城楼纵剖面图
Longitudinal section of gate tower of Great Court Wall

金声玉振坊
Jinsheng yuzhen Archway

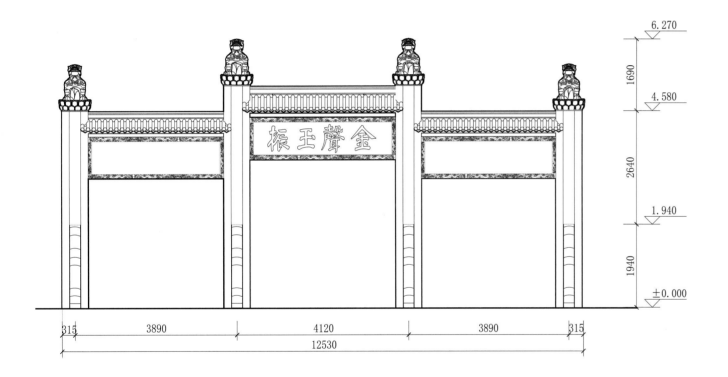

6.270

1690

4.580

2640

1.940

1940

±0.000

315   3890   4120   3890   315

12530

北

金声玉振坊正立面图
Front elevation of Jinsheng yuzhen Archway

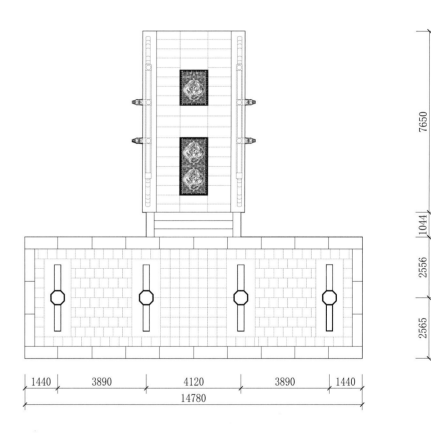

7650

13814

1044

2556

2565

1440   3890   4120   3890   1440

14780

北

金声玉振坊平面图
Plan of Jinsheng yuzhen Archway

中国古建筑测绘大系·祠庙建筑——曲阜孔庙

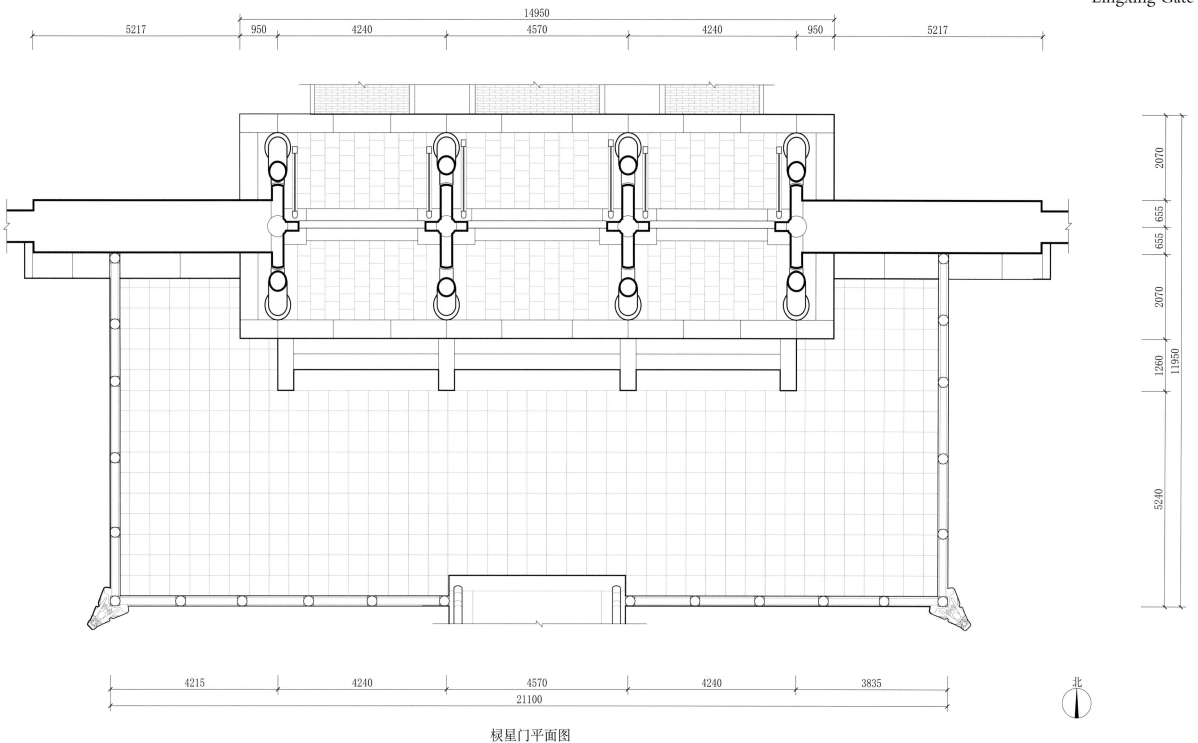

14950

5217　950　4240　4570　4240　950　5217

2070

655

655

2070

1260　11950

5240

030

北

棂星门平面图
Plan of Lingxing Gate

4215　4240　4570　4240　3835

21100

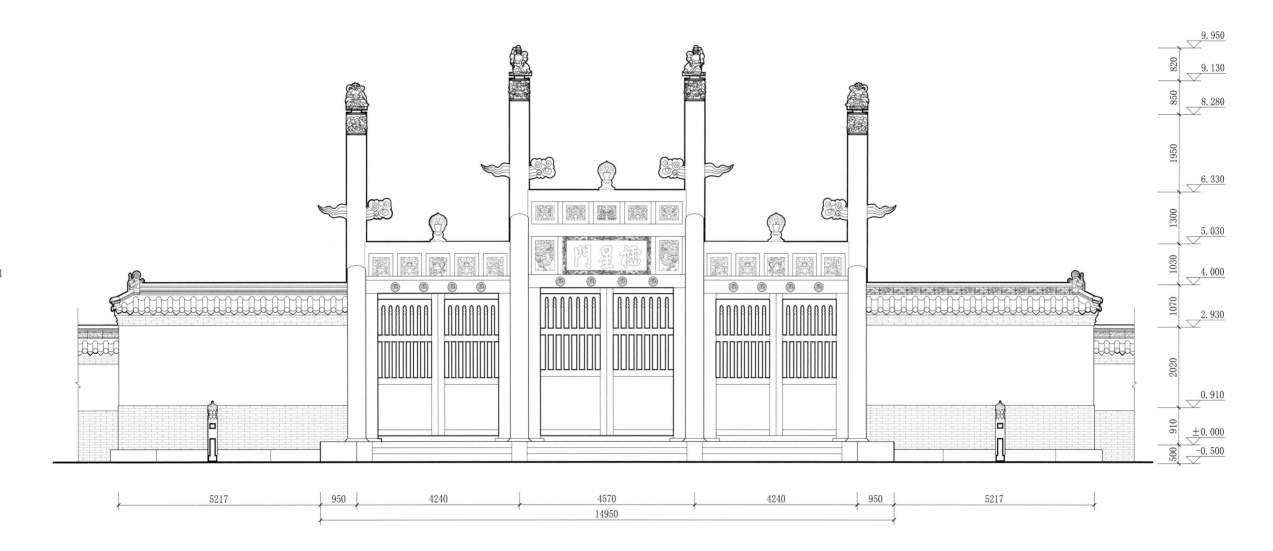

棂星门正立面图
Front elevation of Lingxing Gate

中国古建筑测绘大系·祠庙建筑——曲阜孔庙

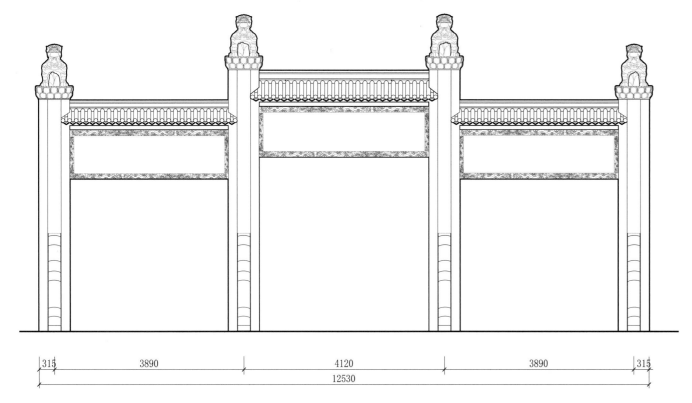

太和元气坊正立面图
Front elevation of Taihe yuanqi Archway

315　　3890　　4120　　3890　　315
12530

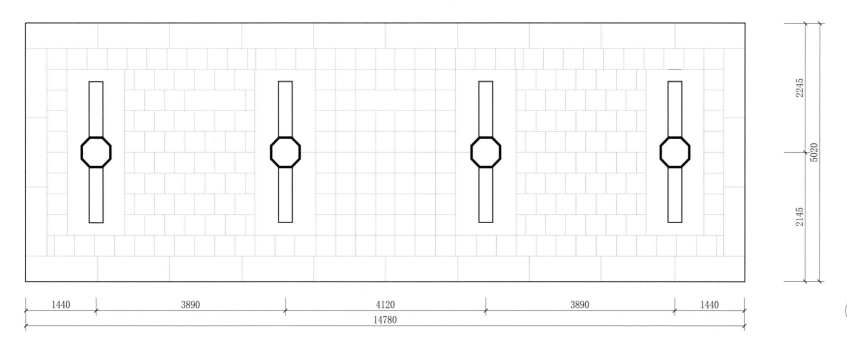

太和元气坊平面图
Plan of Taihe yuanqi Archway

1440　　3890　　4120　　3890　　1440
14780

2245
5020
2145

北

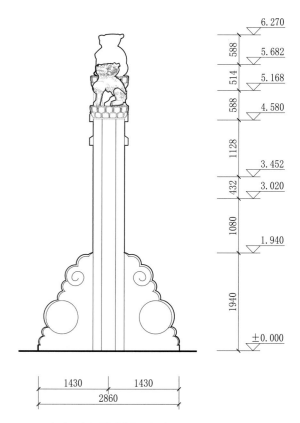

太和元气坊侧立面图
Side elevation of Taihe yuanqi Archway

6.270
588　5.682
514　5.168
588　4.580
1128　3.452
432　3.020
1080　1.940
1940
±0.000

1430　1430
2860

至圣庙坊
Zhisheng Archway

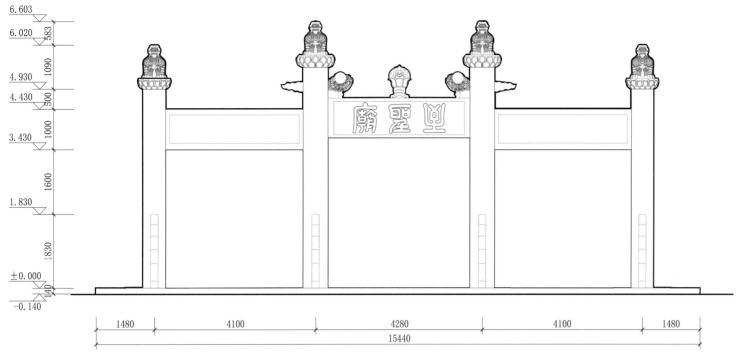

至圣庙坊正立面图
Front elevation of Zhisheng Archway

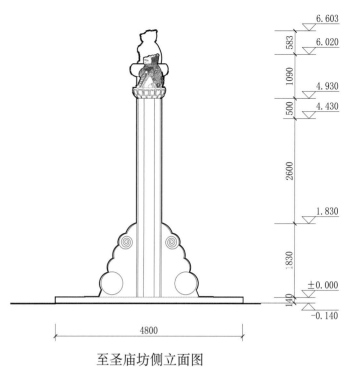

至圣庙坊侧立面图
Side elevation of Zhisheng Archway

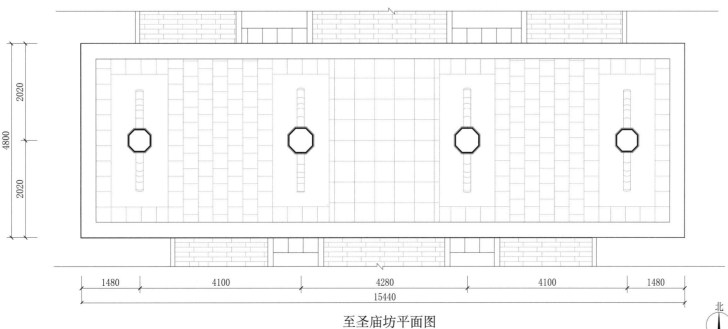

至圣庙坊平面图
Plan of Zhisheng Archway

北

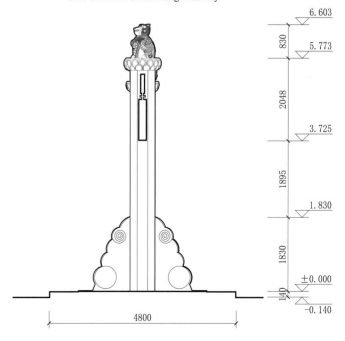

至圣庙坊明间剖面图
Section of central bay of Zhisheng Archway

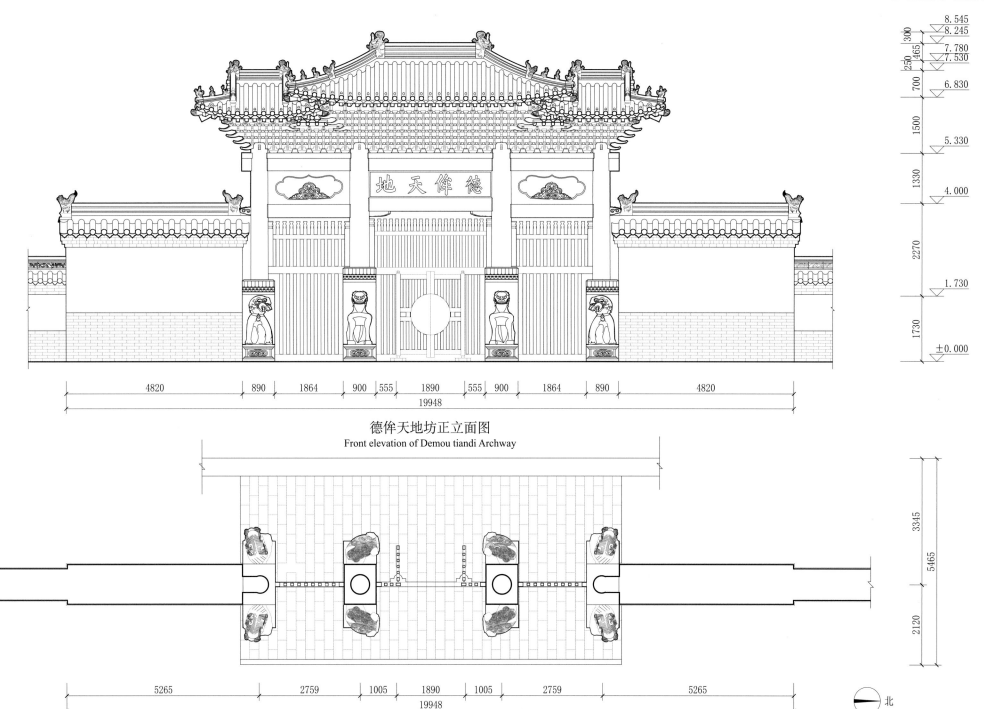

8.545
8.245
300
7.780
250 465
7.530
700
6.830
1500
5.330
1330
4.000
2270
1.730
1730
±0.000

4820 | 890 | 1864 | 900 | 555 | 1890 | 555 | 900 | 1864 | 890 | 4820
19948

德侔天地坊正立面图
Front elevation of Demou tiandi Archway

3345
5465
2120

5265 | 2759 | 1005 | 1890 | 1005 | 2759 | 5265
19948

德侔天地坊平面图
Plan of Demou tiandi Archway

北

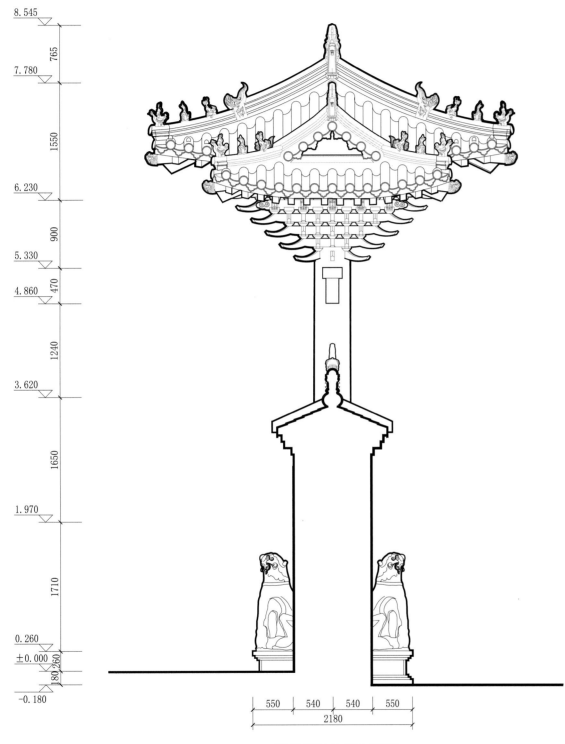

德侔天地坊侧立面图
Side elevation of Demou tiandi Archway

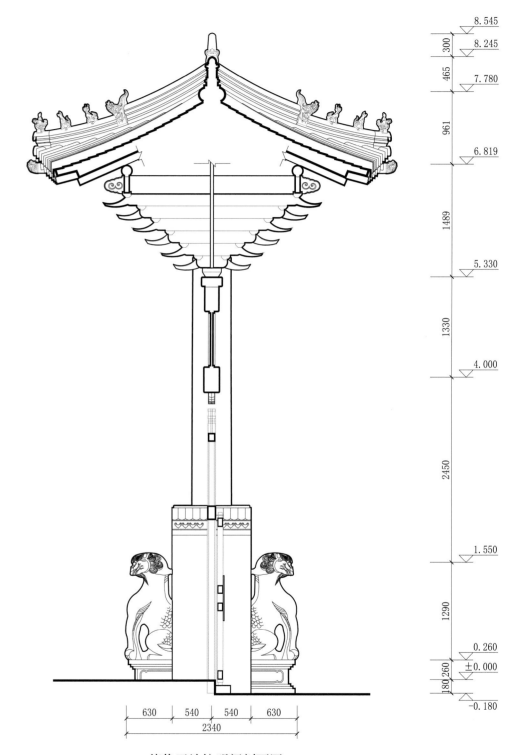

德侔天地坊明间剖面图
Section of central bay of Demou tiandi Archway

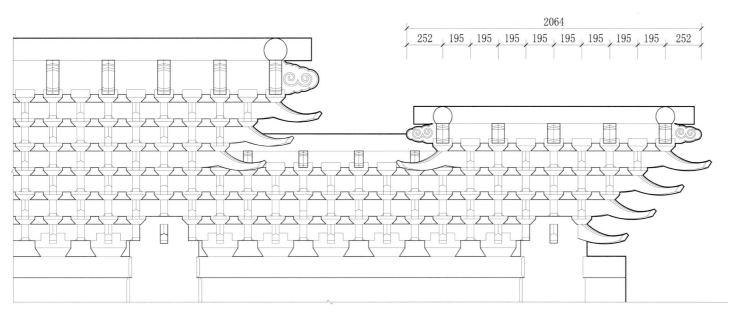

正立面图
Front view

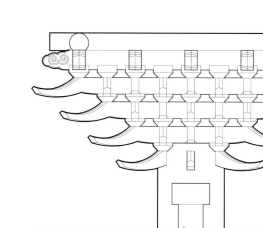

侧立面图
Side view

036

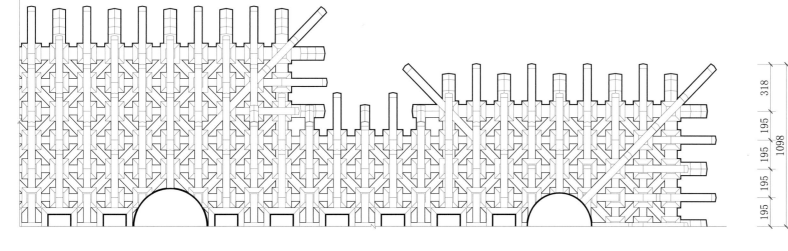

仰视图
View from below

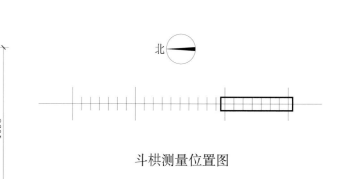

北

斗栱测量位置图

德侔天地坊斗栱大样图
Brackets of Demou tiandi Archway

圣时门
Shengshi Gate

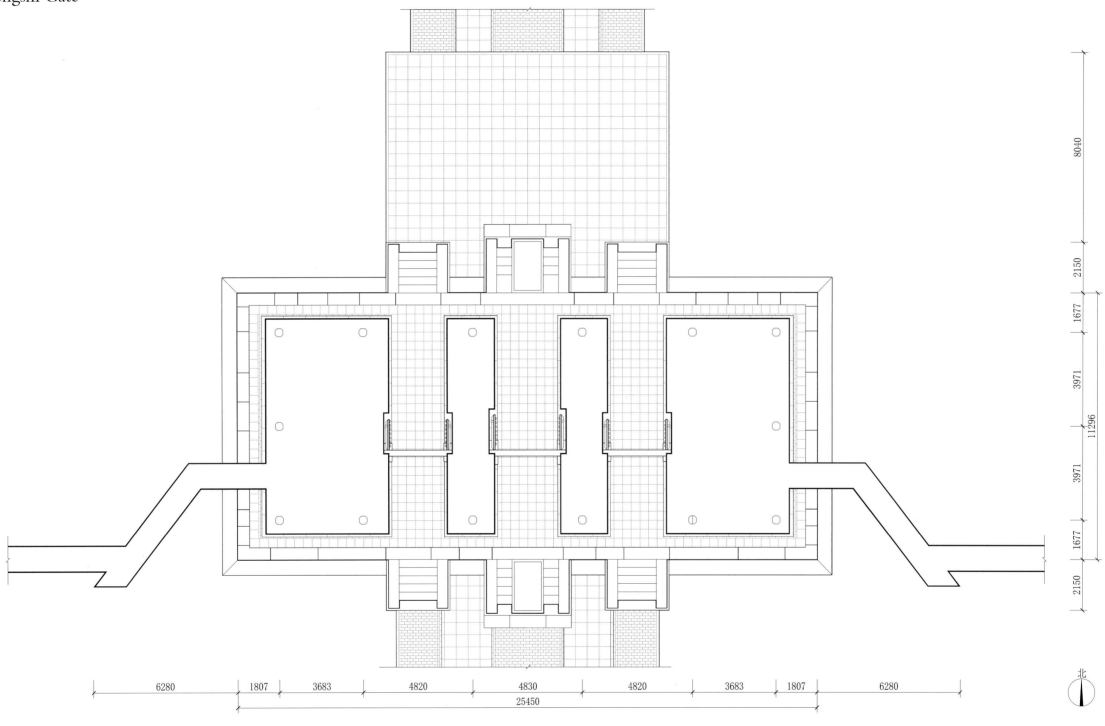

圣时门平面图
Plan of Shengshi Gate

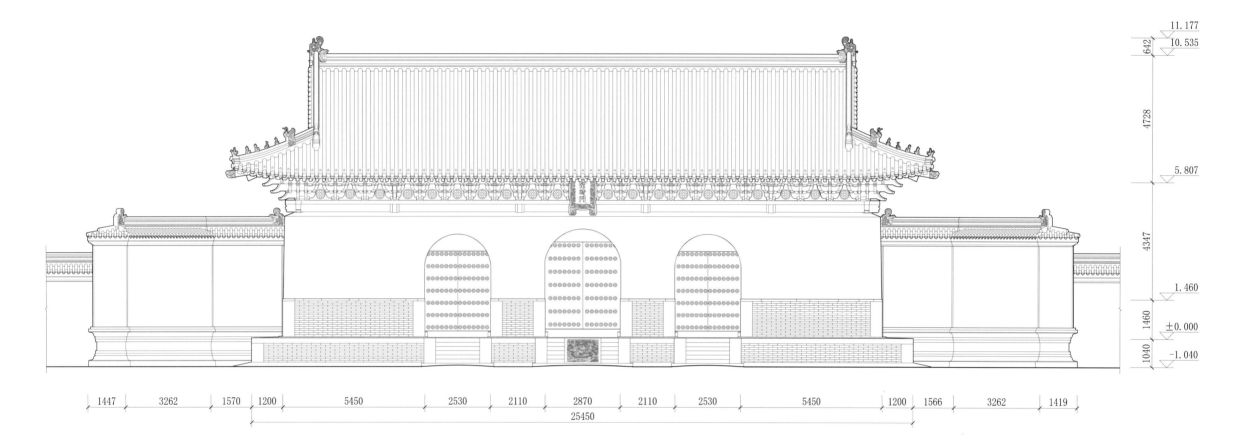

11.177
642
10.535
4728
5.807
4347
1.460
1460
±0.000
1040
−1.040

1447　3262　1570　1200　5450　2530　2110　2870　2110　2530　5450　1200　1566　3262　1419

25450

圣时门正立面图
Front elevation of Shengshi Gate

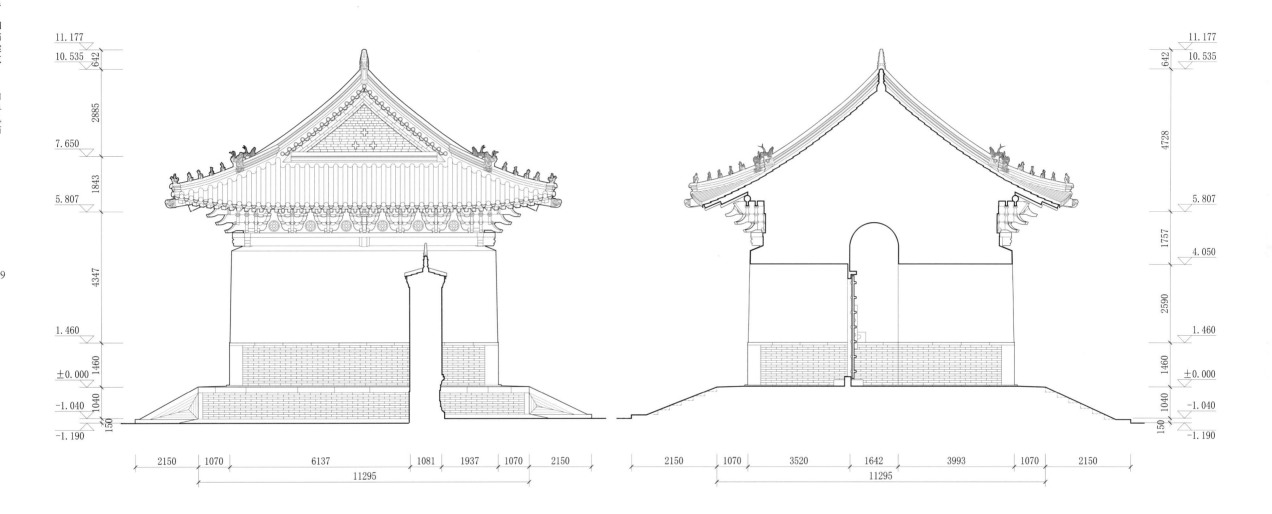

圣时门侧立面图
Side elevation of Shengshi Gate

圣时门明间剖面图
Section of central bay of Shengshi Gate

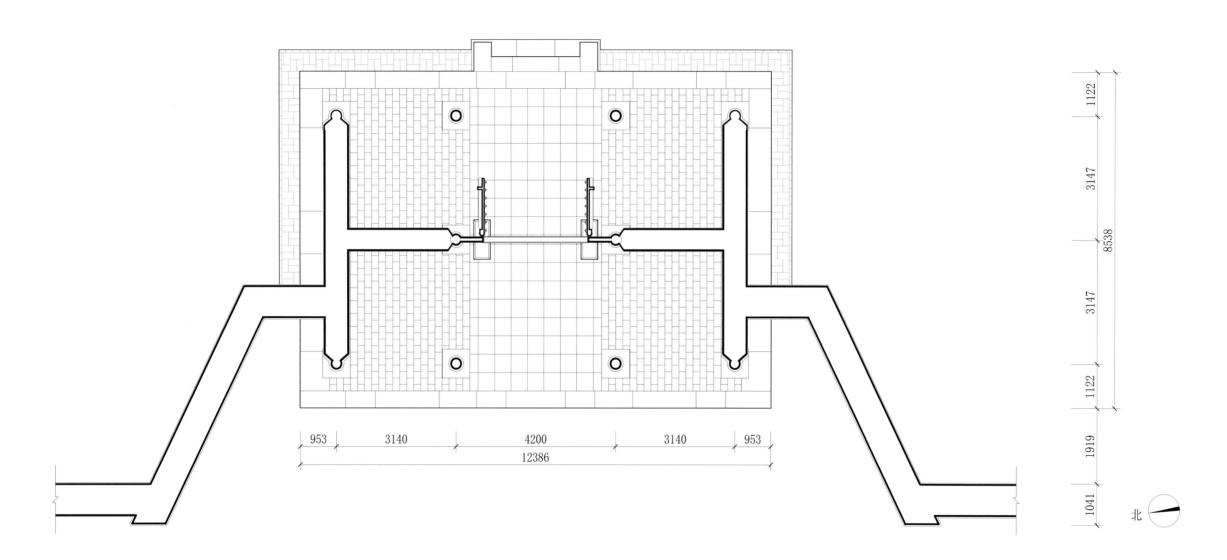

1122
3147
8538
3147
1122
1919
1041

953 3140 4200 3140 953
12386

北

仰高门平面图
Plan of Yanggao Gate

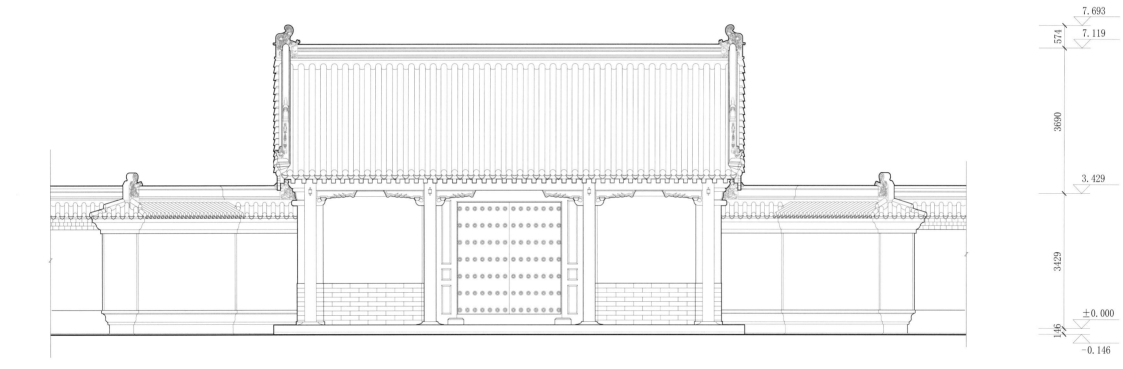

仰高门正立面图
Front elevation of Yanggao Gate

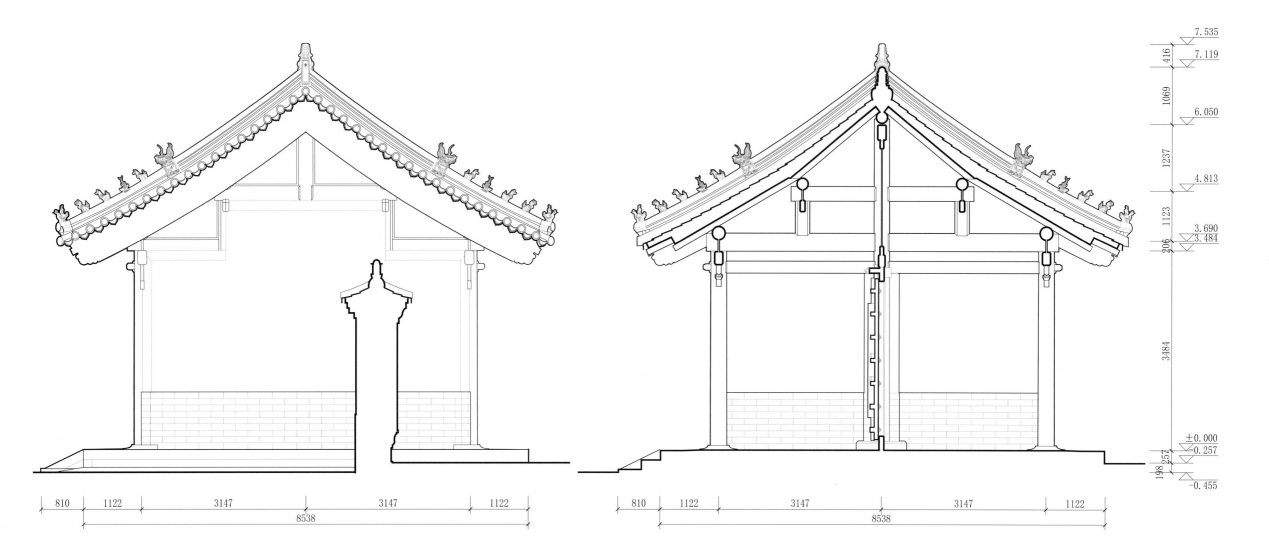

仰高门侧立面图
Side elevation of Yanggao Gate

仰高门横剖面图
Cross-section of Yanggao Gate

璧水桥
Bishui Bridge

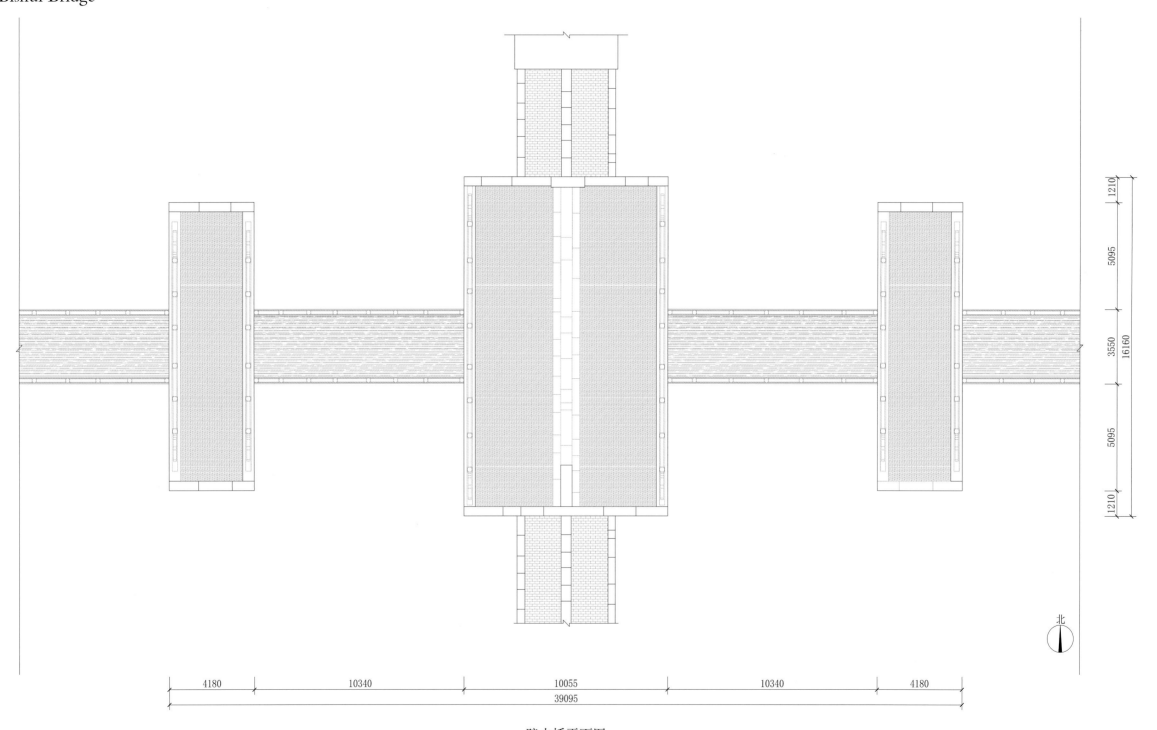

北

璧水桥平面图
Plan of Bishui Bridge

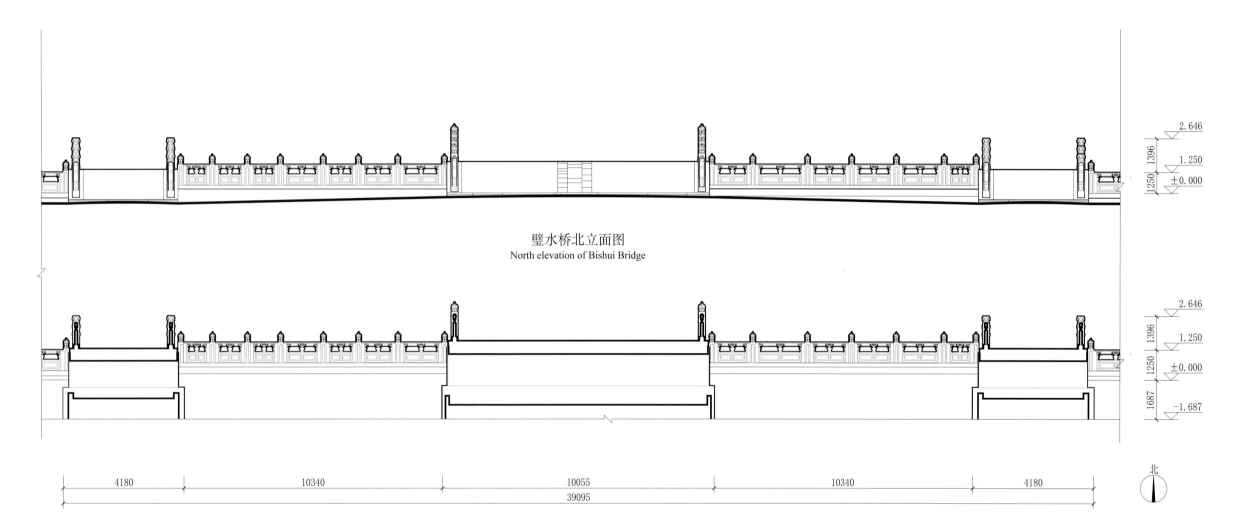

2.646

1396

1250 | 1.250

±0.000

璧水桥北立面图
North elevation of Bishui Bridge

2.646

1396

1250 | 1.250

1687 | ±0.000

−1.687

4180　10340　10055　10340　4180

39095

北

璧水桥纵剖面图
Longitudinal section of Bishui Bridge

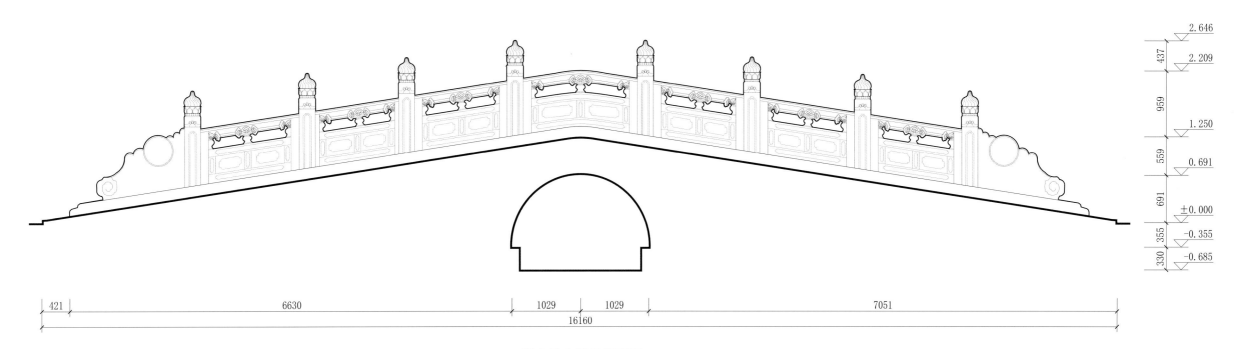

璧水桥中路纵剖面图
Longitudinal section of central path of Bishui Bridge

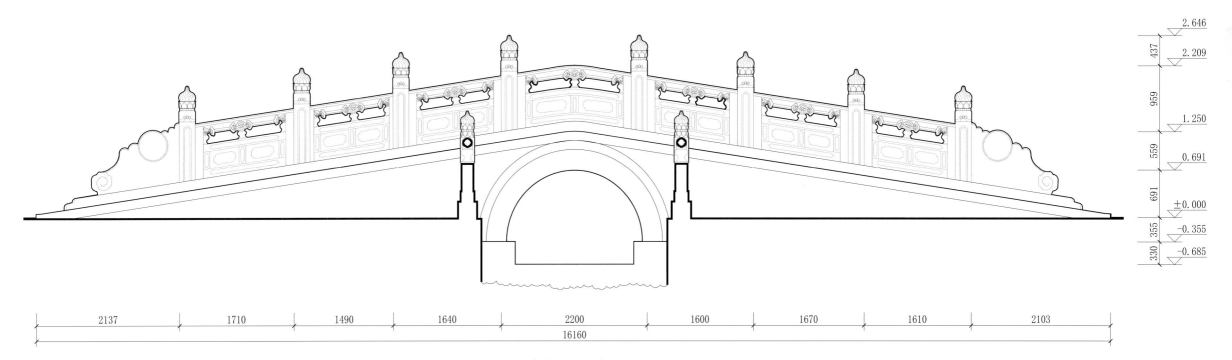

璧水桥中路西立面图
West elevation of Bishui Bridge

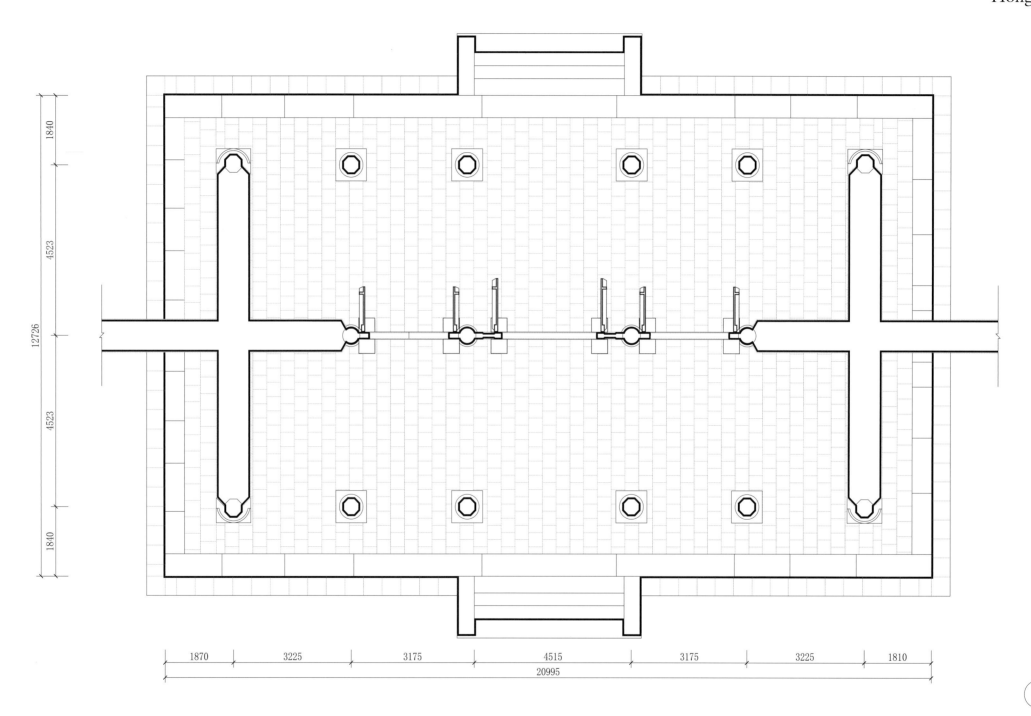

弘道门平面图
Plan of Hongdao Gate

北

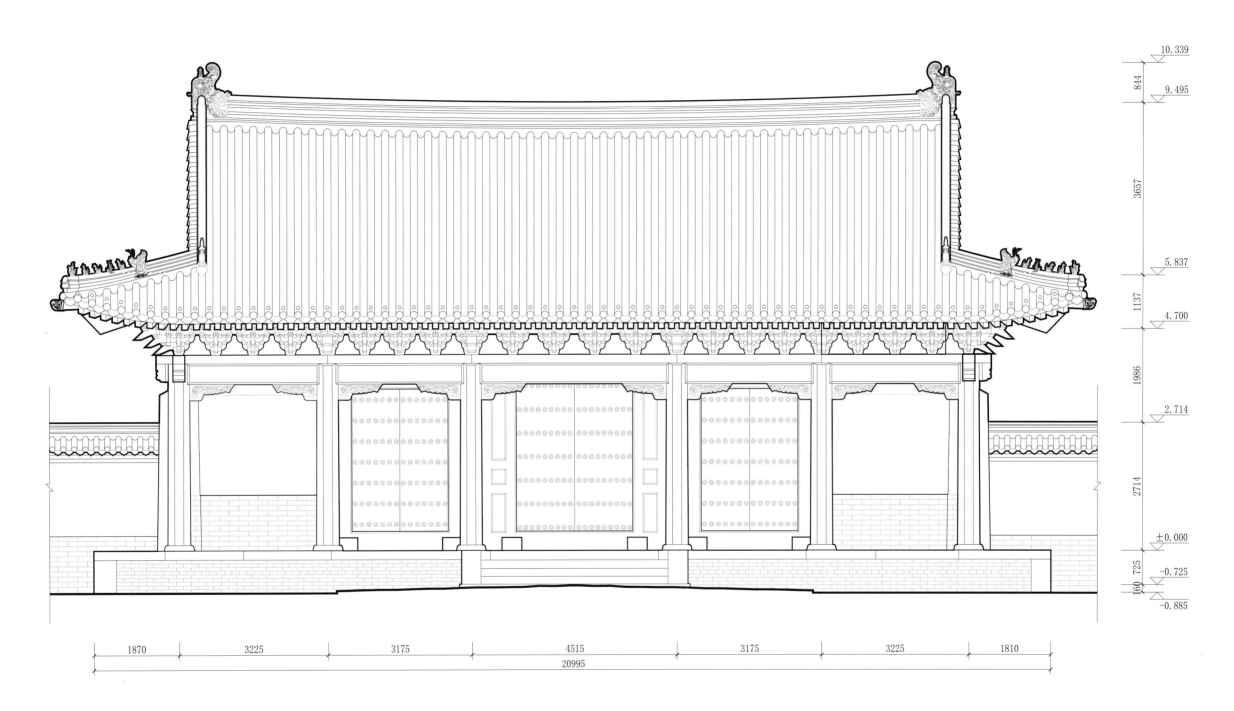

10.339

844

9.495

3657

5.837

1137

4.700

1986

2.714

2714

±0.000

725

160

-0.725

-0.885

1870　3225　3175　4515　3175　3225　1810

20995

弘道门正立面图
Front elevation of Hongdao Gate

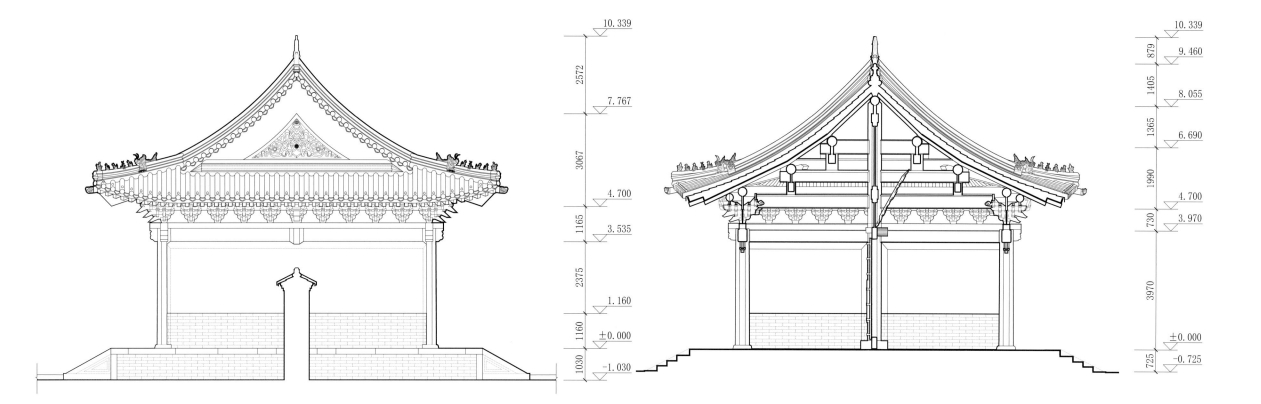

10.339

2572

7.767

3067

4.700

1165

3.535

2375

1.160

1160

±0.000

1030

-1.030

10.339

879   9.460

1405   8.055

1365   6.690

1990   4.700

730   3.970

3970

±0.000

725   -0.725

1630 | 1840 | 4523 | 4523 | 1840 | 1630

12726

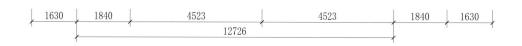

1630 | 1840 | 4523 | 4523 | 1840 | 1630

12726

弘道门侧立面图
Side elevation of Hongdao Gate

弘道门明间剖面图
Section of central bay of Hongdao Gate

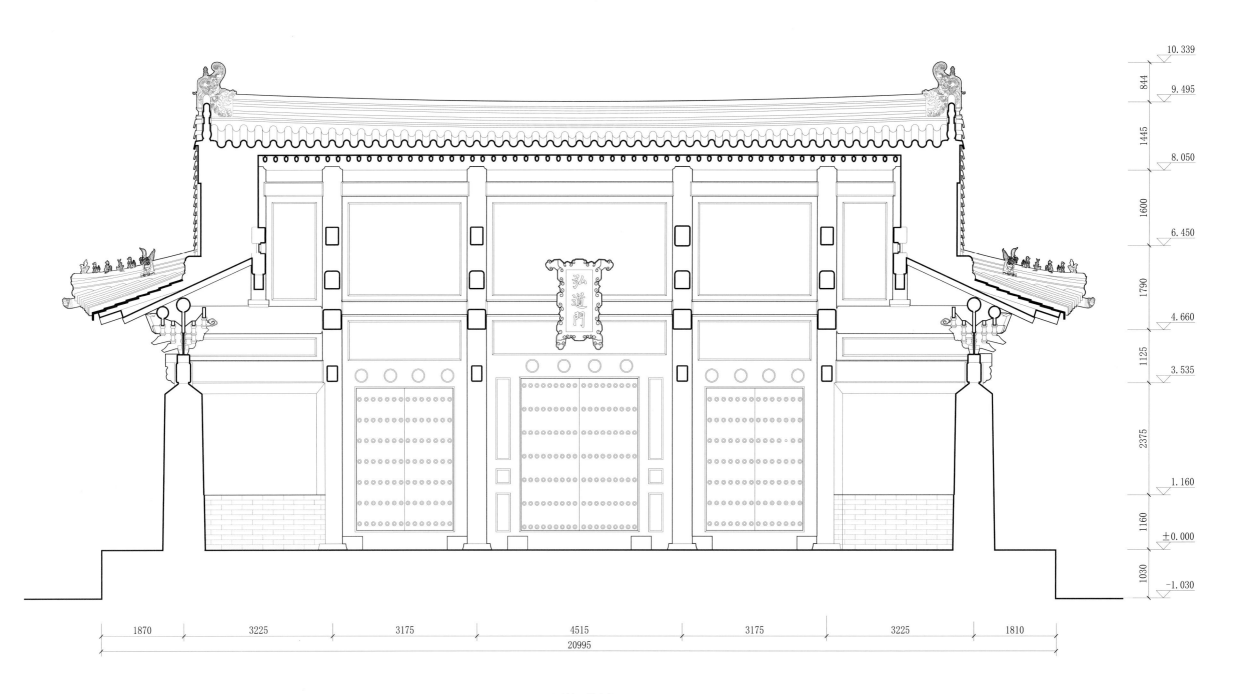

10.339
844
9.495
1445
8.050
1600
6.450
1790
4.660
1125
3.535
2375
1.160
1160
±0.000
1030
-1.030

1870   3225   3175   4515   3175   3225   1810
20995

弘道门纵剖面图
Longitudinal section of Hongdao Gate

奎文阁和大成门前院

Kuiwen Pavilion and
the front courtyard of
Dacheng Gate

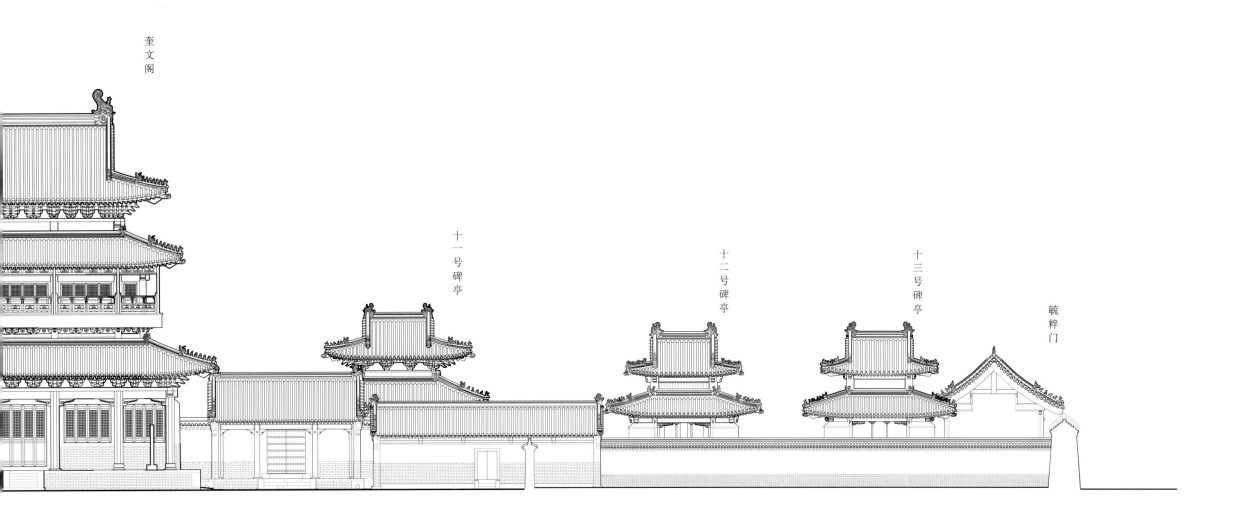

奎文阁

十一号碑亭

十二号碑亭

十三号碑亭

毓粹门

0　　3m

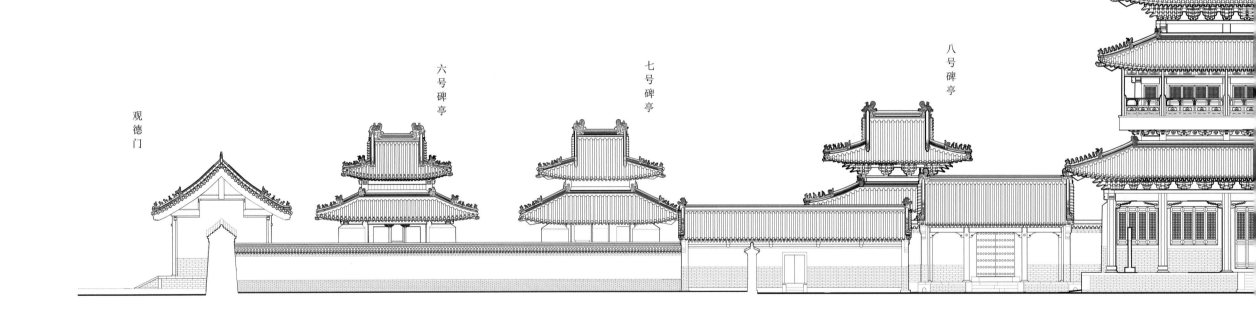

观德门

六号碑亭

七号碑亭

八号碑亭

奎文阁群体立面图
Elevation of Kuiwen Pavilion building complex

中国古建筑测绘大系·祠庙建筑——曲阜孔庙

054

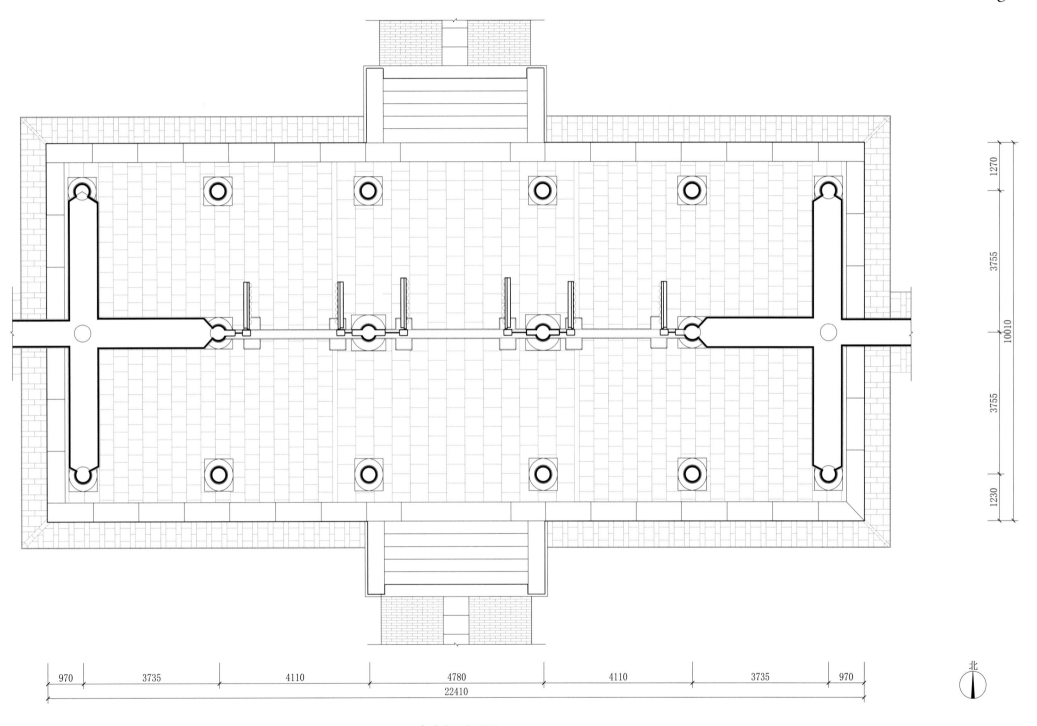

1270

3755

10010

3755

1230

970　3735　4110　4780　4110　3735　970

22410

北

大中门平面图
Plan of Dazhong Gate

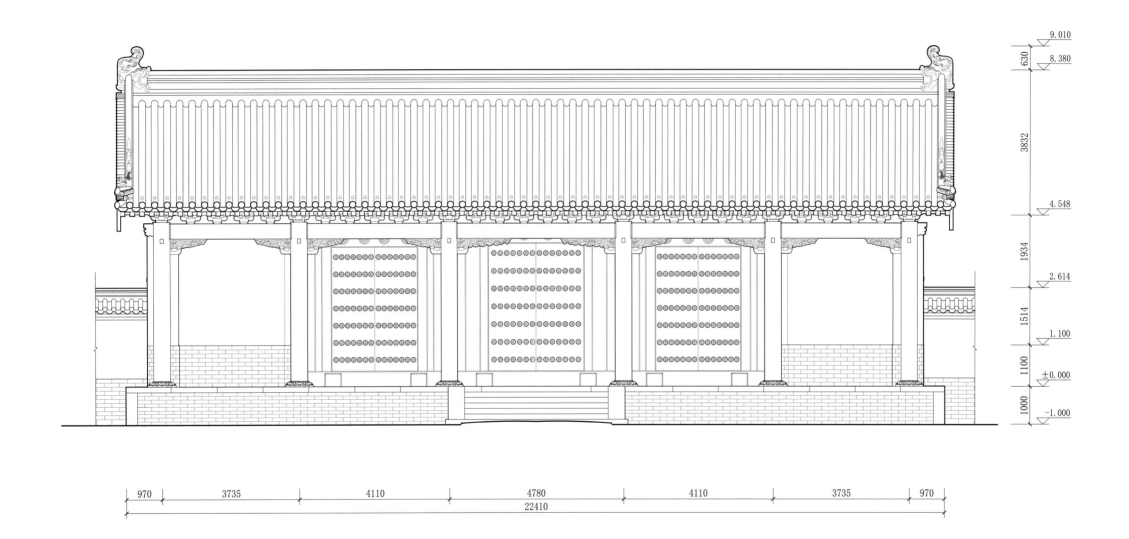

9.010
630
8.380

3832

4.548

1934

2.614

1514

1.100

1100

±0.000

1000

-1.000

970   3735   4110   4780   4110   3735   970

22410

大中门正立面图
Front elevation of Dazhong Gate

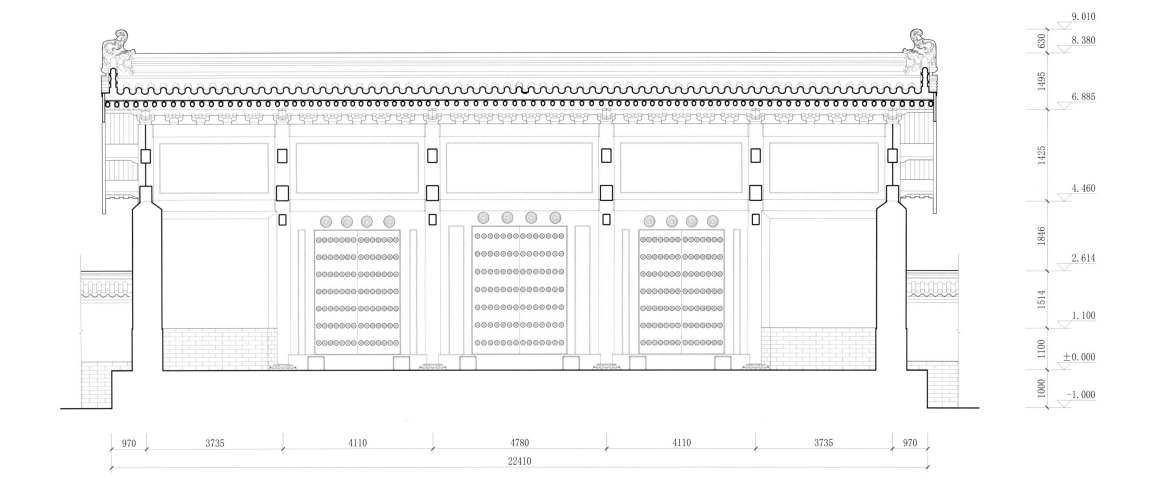

9.010
8.380
6.885
4.460
2.614
1.100
±0.000
-1.000

630
1495
1425
1846
1514
1100
1000

970　3735　4110　4780　4110　3735　970

22410

大中门纵剖面图
Longitudinal section of Dazhong Gate

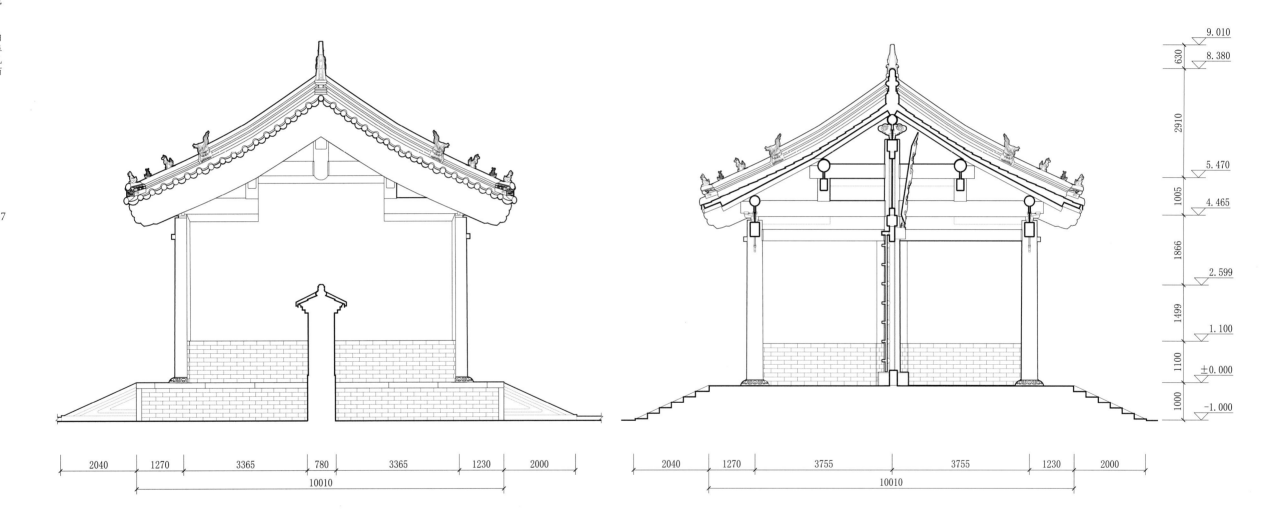

大中门侧立面图
Side elevation of Dazhong Gate

大中门明间剖面图
Section of central bay of Dazhong Gate

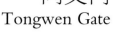

中国古建筑测绘大系·祠庙建筑——曲阜孔庙

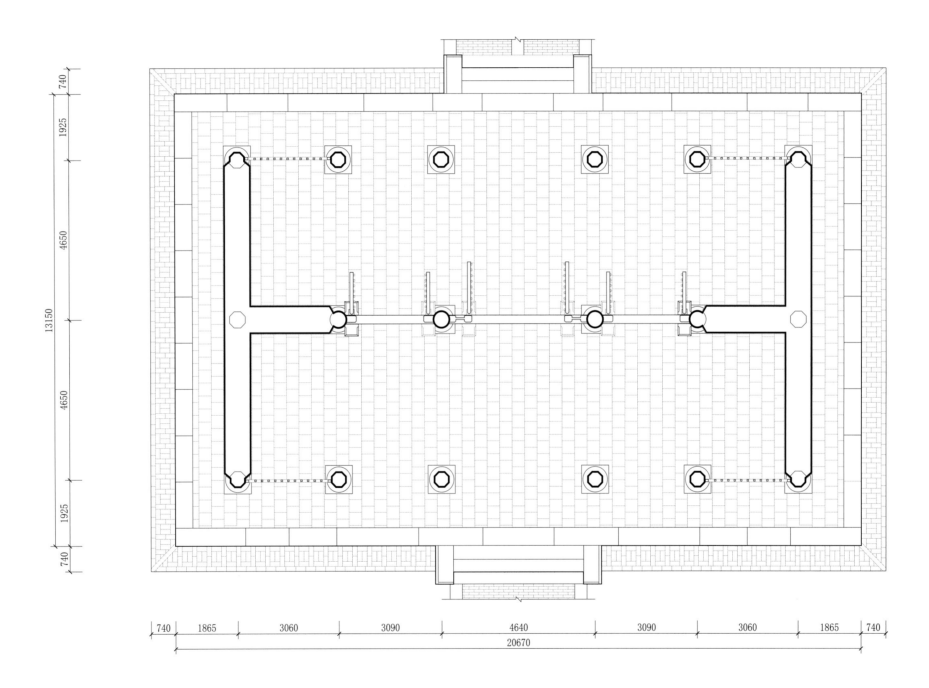

北

同文门平面图
Plan of Tongwen Gate

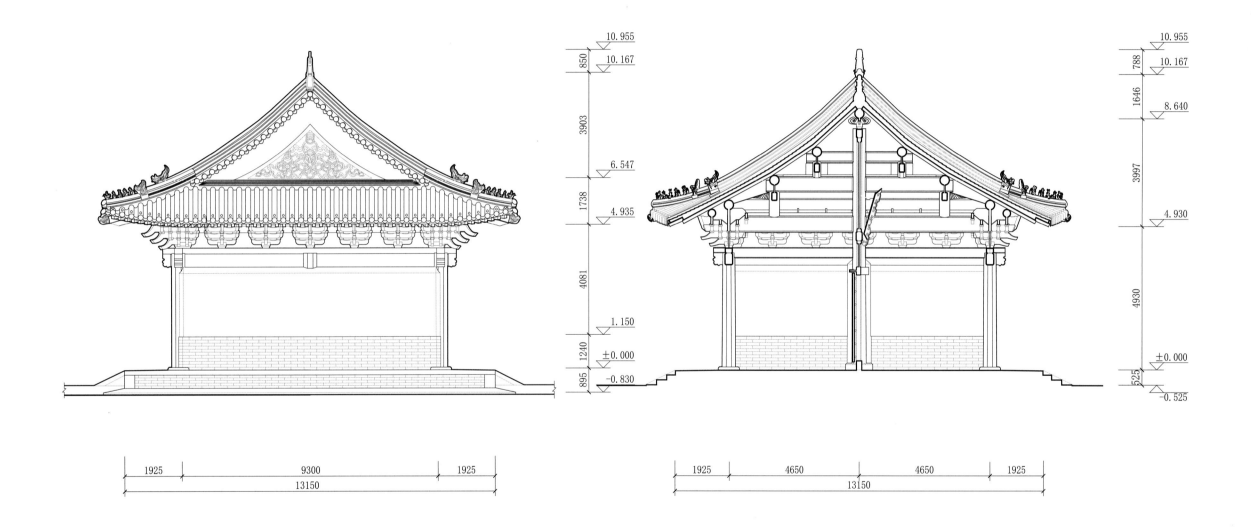

10.955
850
10.167
3903
6.547
1738
4.935
4081
1.150
1240
±0.000
895
-0.830

10.955
788
10.167
1646
8.640
3997
4.930
4930
±0.000
525
-0.525

1925 9300 1925
13150

1925 4650 4650 1925
13150

同文门侧立面图
Side elevation of Tongwen Gate

同文门明间剖面图
Section of central bay of Tongwen Gate

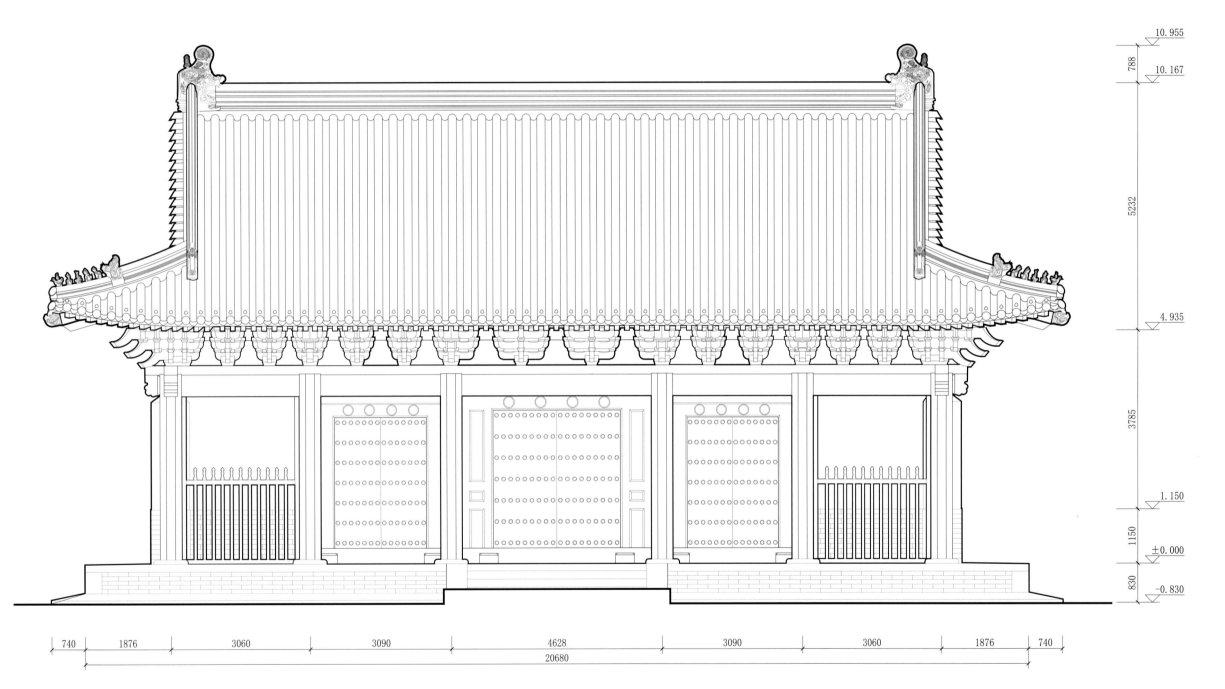

同文门正立面图
Front elevation of Tongwen Gate

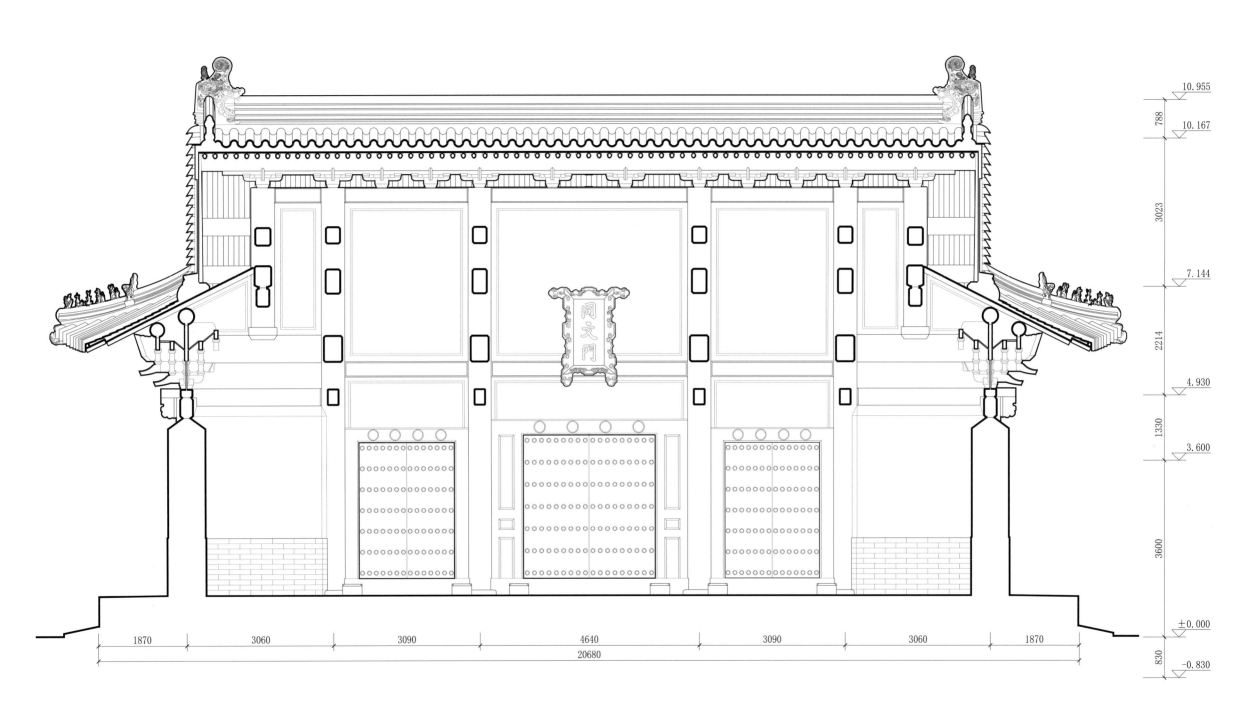

10.955

788

10.167

3023

7.144

2214

4.930

1330

3.600

3600

±0.000

830

−0.830

| 1870 | 3060 | 3090 | 4640 | 3090 | 3060 | 1870 |

20680

同文门纵剖面图
Longitudinal section of Tongwen Gate

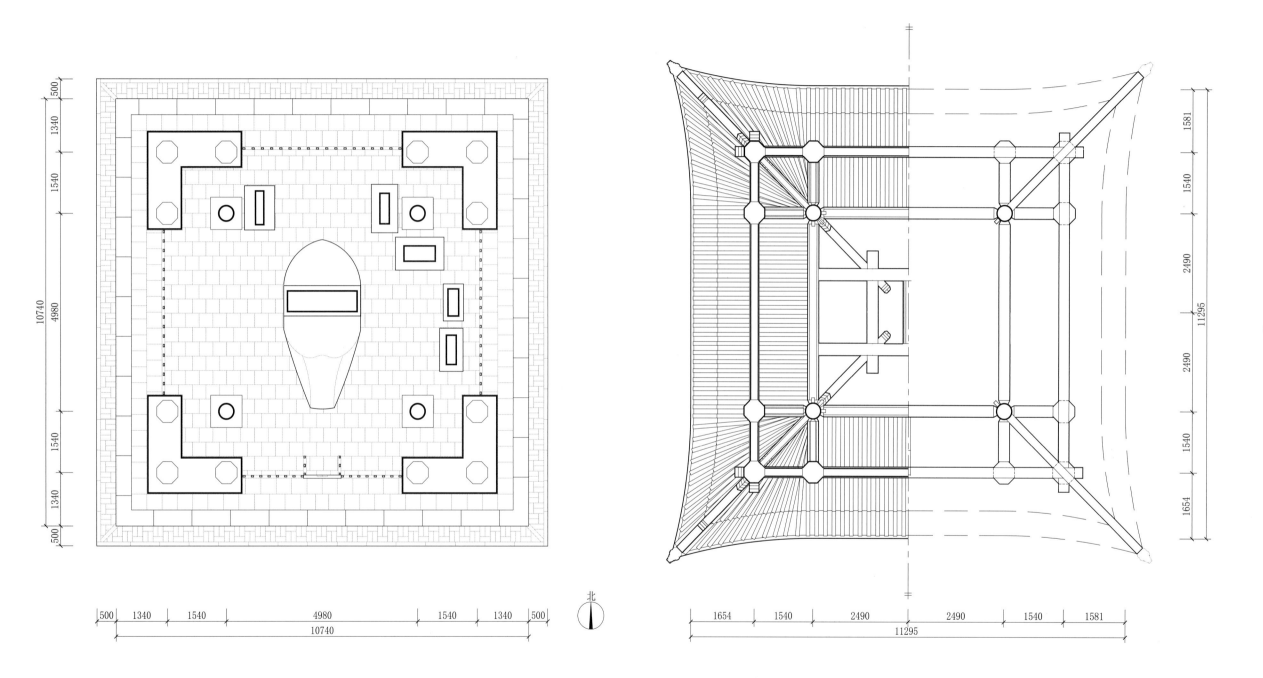

明洪武碑东碑亭平面图
Plan of east stele pavilion of Hongwu period, Ming Dynasty

明洪武碑东碑亭一层梁架仰俯视图
Plan of first-floor framework of east stele pavilion of Hongwu period, Ming Dynasty, as seen from below

北

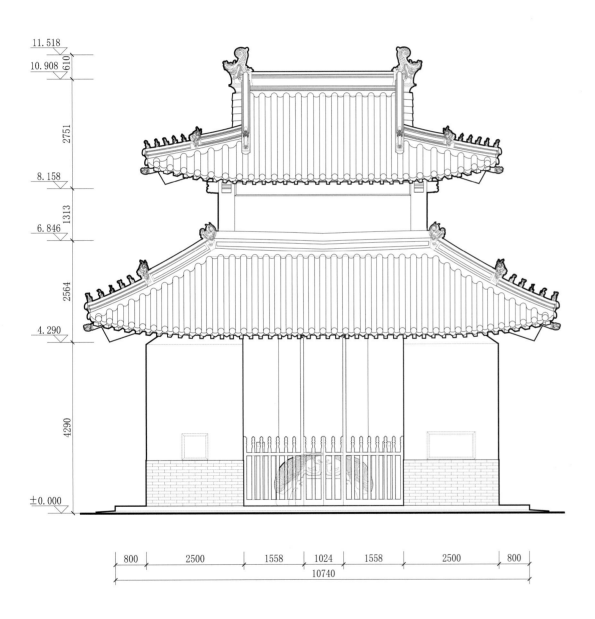

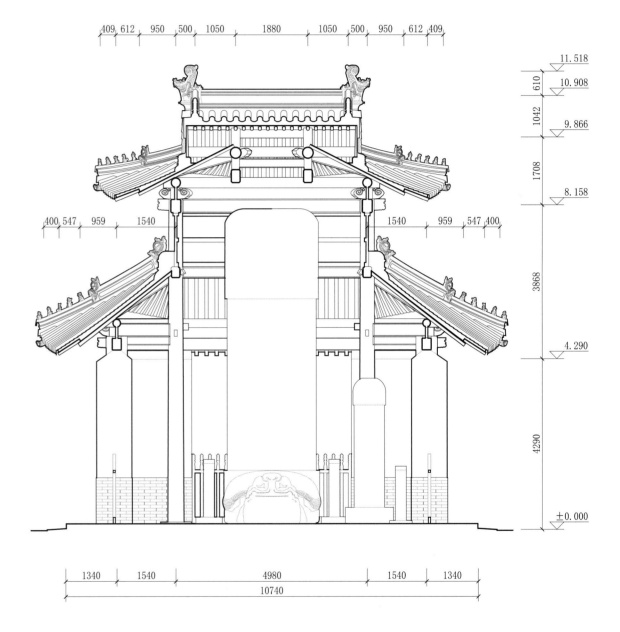

明洪武碑东碑亭正立面图
Frontal elevation of east stele pavilion of Hongwu period, Ming Dynasty

明洪武碑东碑亭纵剖面图
Longitudinal section of east stele pavilion of Hongwu period, Ming Dynasty

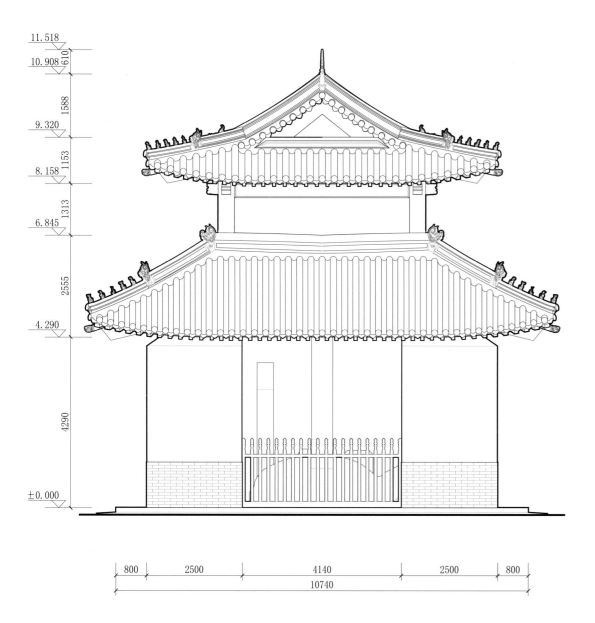

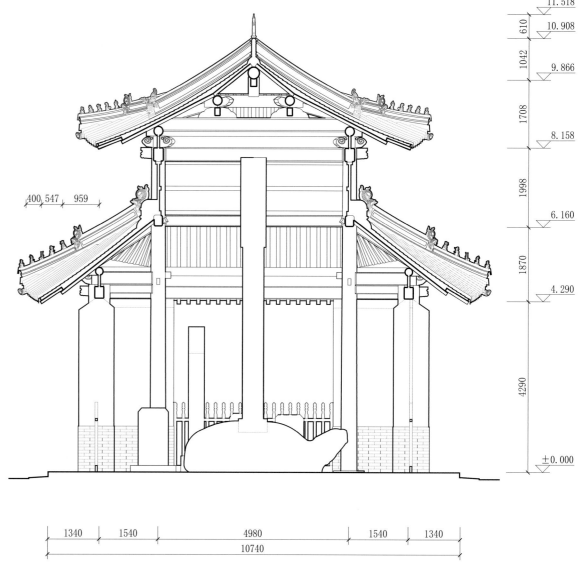

明洪武碑东碑亭侧立面图
Side elevation of east stele pavilion of Hongwu period, Ming Dynasty

明洪武碑东碑亭横剖面图
Cross-section of east stele pavilion of Hongwu period, Ming Dynasty

明洪武碑西碑亭
West stele pavilion of Hongwu period, Ming Dynasty

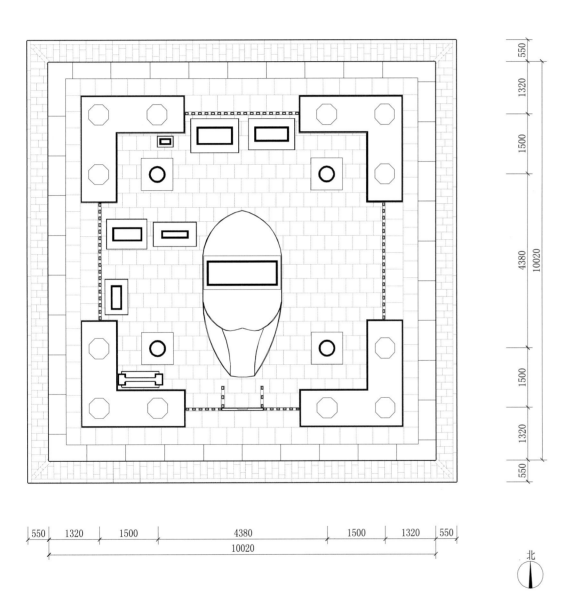

明洪武碑西碑亭平面图
Plan of west stele pavilion of Hongwu period, Ming Dynasty

北

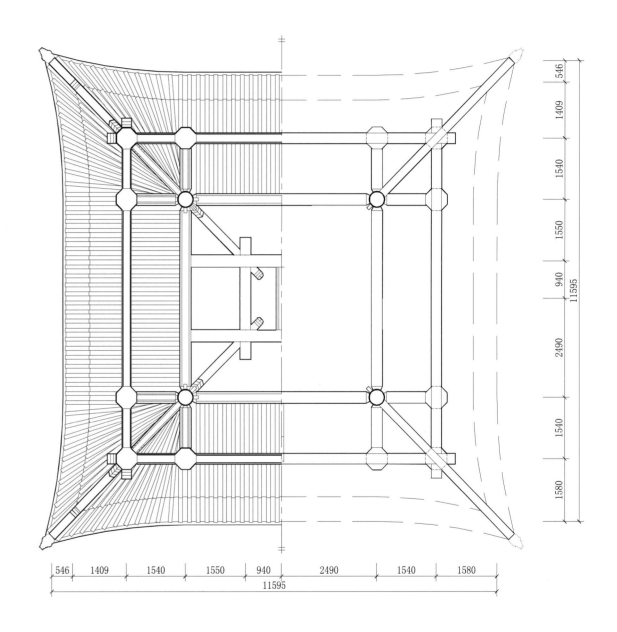

明洪武碑西碑亭一层梁架仰俯视图
Plan of first-floor framework of west stele pavilion of Hongwu period, Ming Dynasty, as seen from below

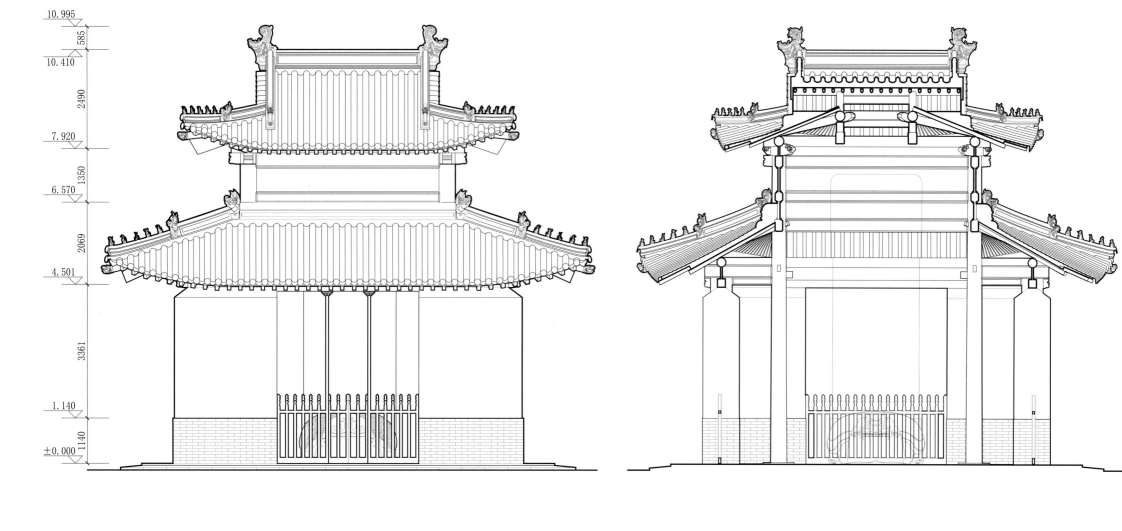

明洪武碑西碑亭正立面图
Front elevation of west stele pavilion of Hongwu period, Ming Dynasty

明洪武碑西碑亭纵剖面图
Longitudinal section of west stele pavilion of Hongwu period, Ming Dynasty

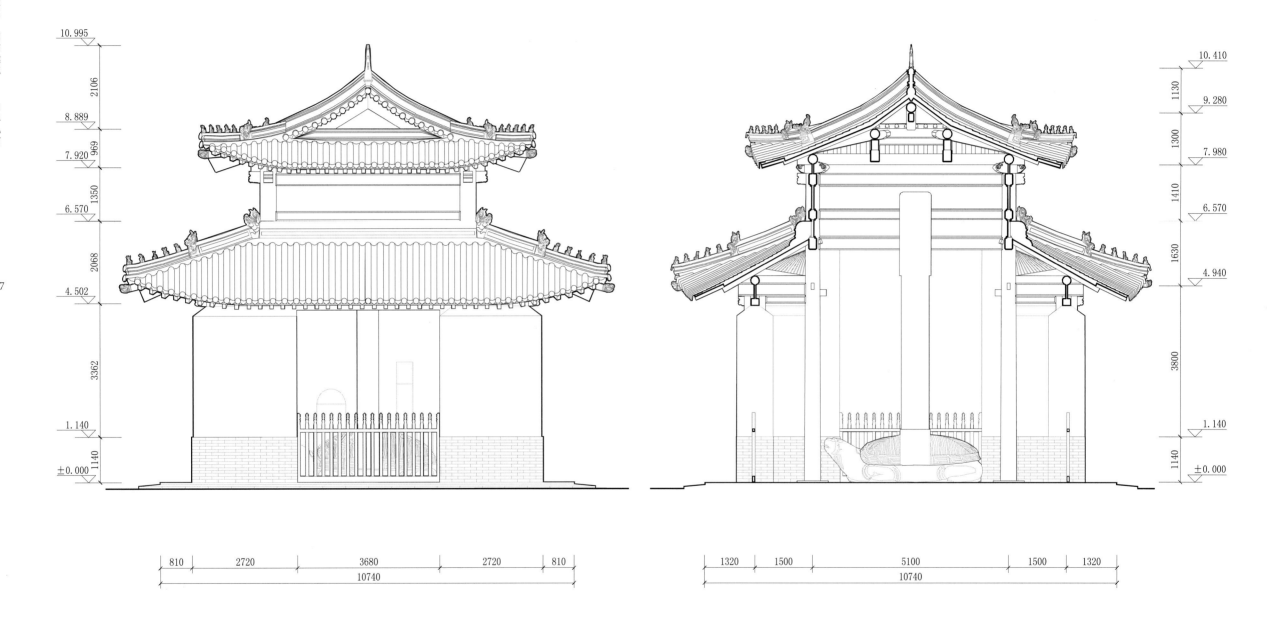

明洪武碑西碑亭侧立面图
Side elevation of west stele pavilion of Hongwu period, Ming Dynasty

明洪武碑西碑亭横剖面图
Cross-section of west stele pavilion of Hongwu period, Ming Dynasty

中国古建筑测绘大系·祠庙建筑——曲阜孔庙

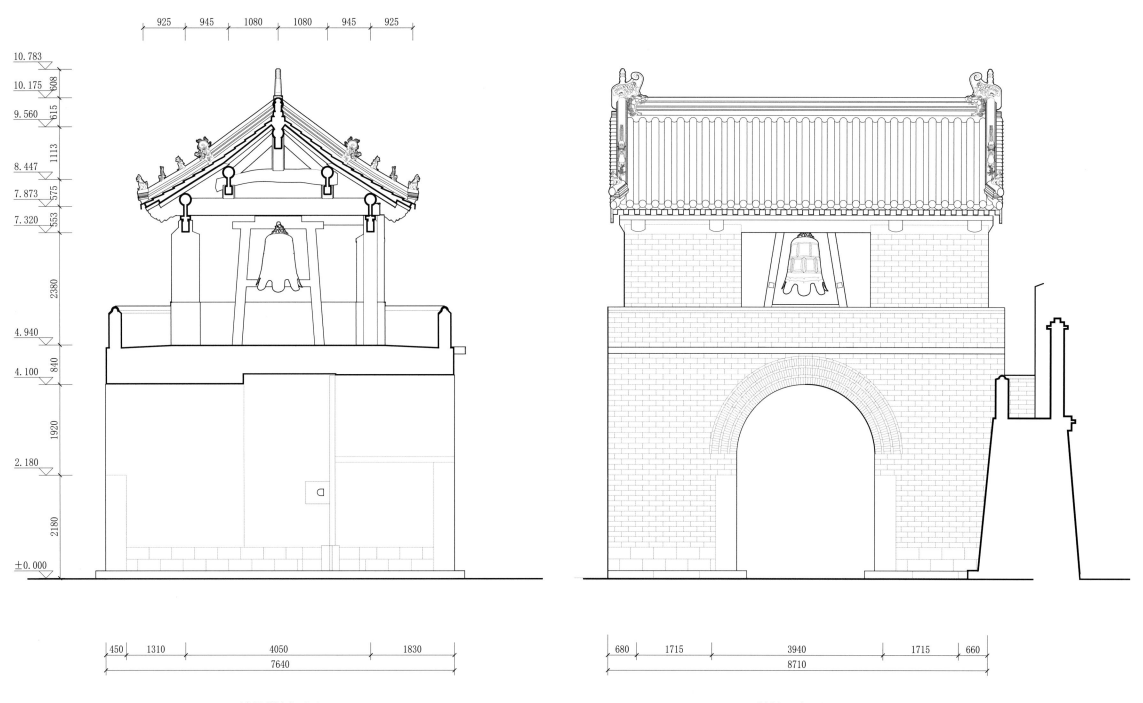

钟楼横剖面图
Cross-section of bell tower

钟楼正立面图
Front elevation of bell tower

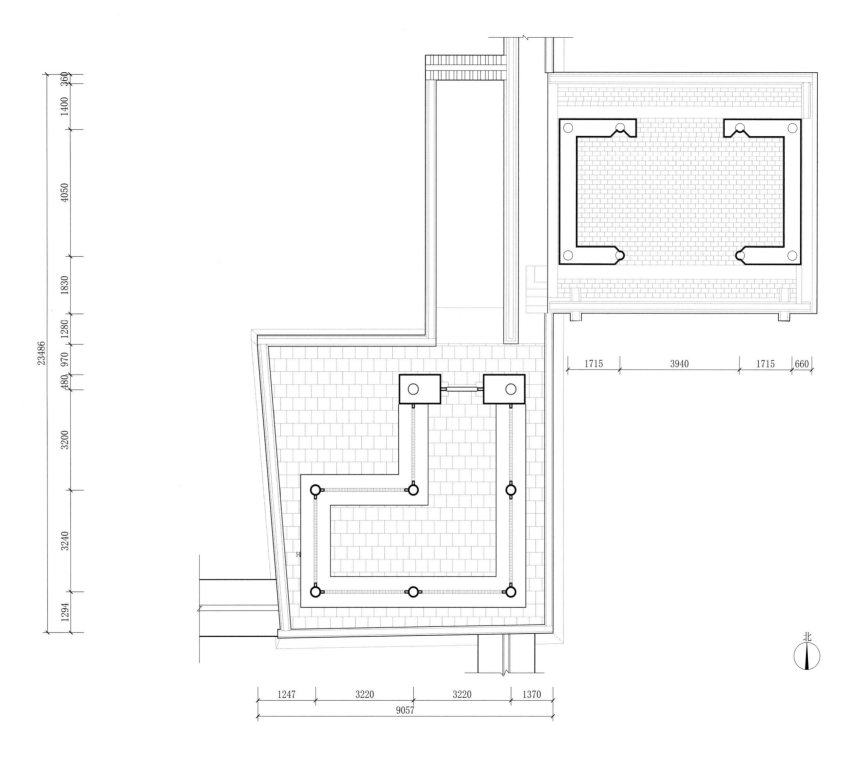

东南角楼与钟楼平面图
Plan of southeast corner building and bell tower

北

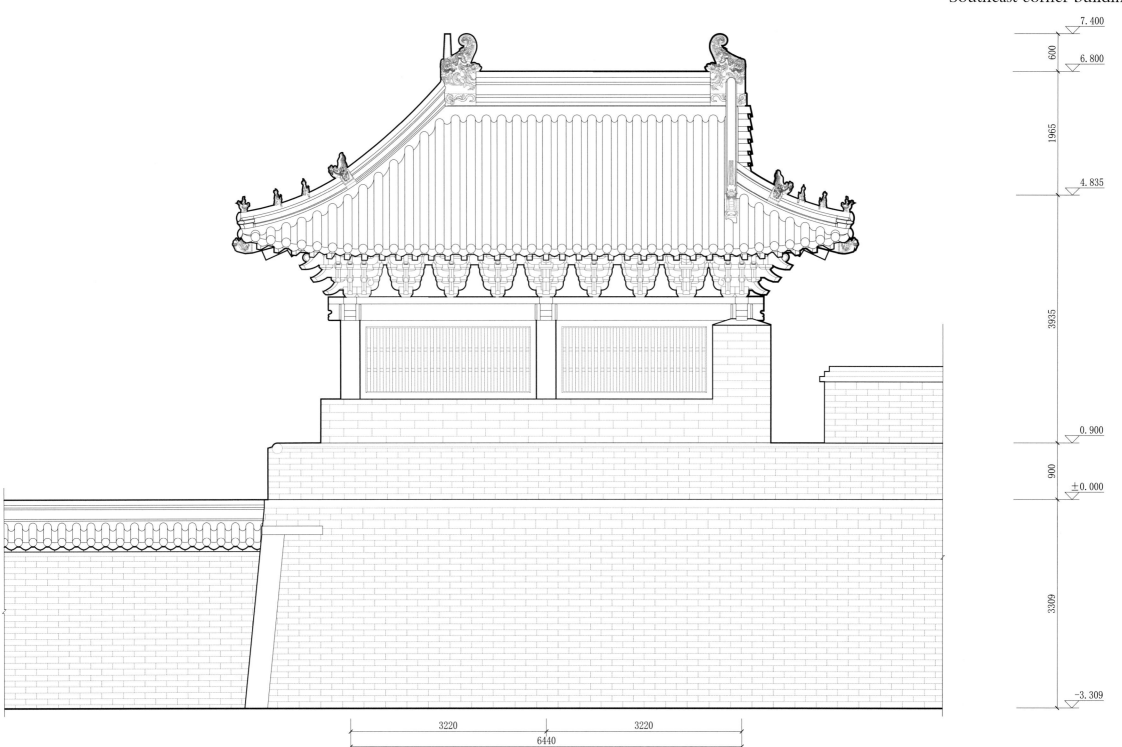

7.400

600

6.800

1965

4.835

3935

0.900

900

±0.000

3309

-3.309

3220　　3220

6440

东南角楼东立面图

East elevation of southeast corner building

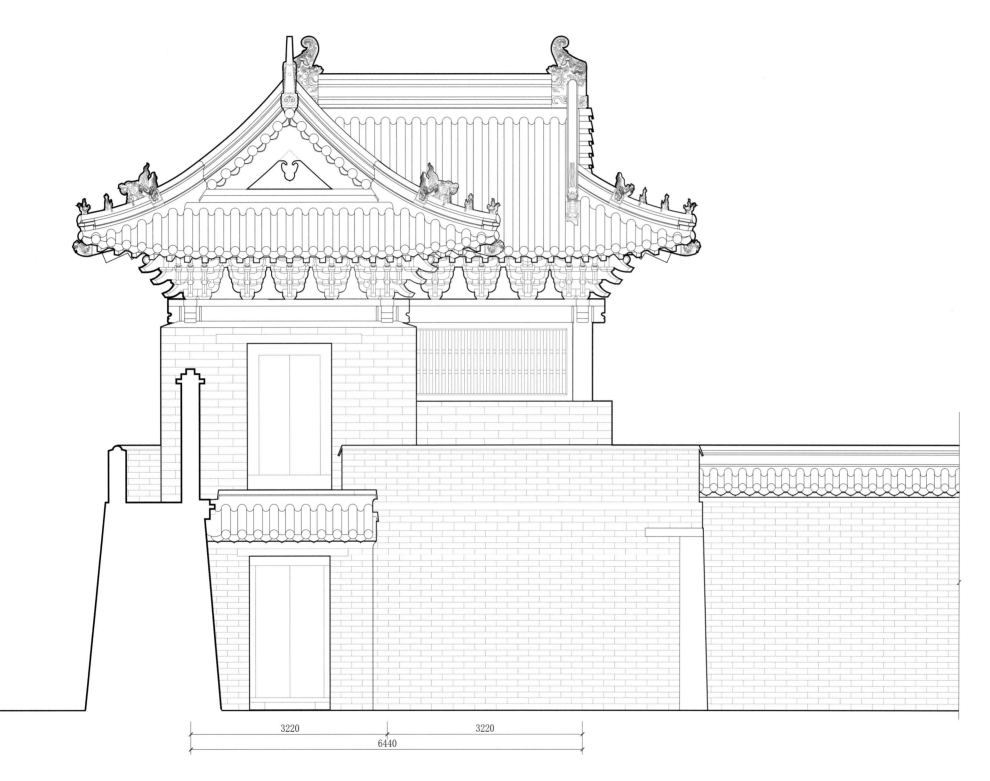

7.400

600

6.800

2913

3.887

777

3.110

1510

1.600

700

900

900

±0.000

3310

-3.310

3220

3220

6440

东南角楼北立面图
North elevation of southeast corner building

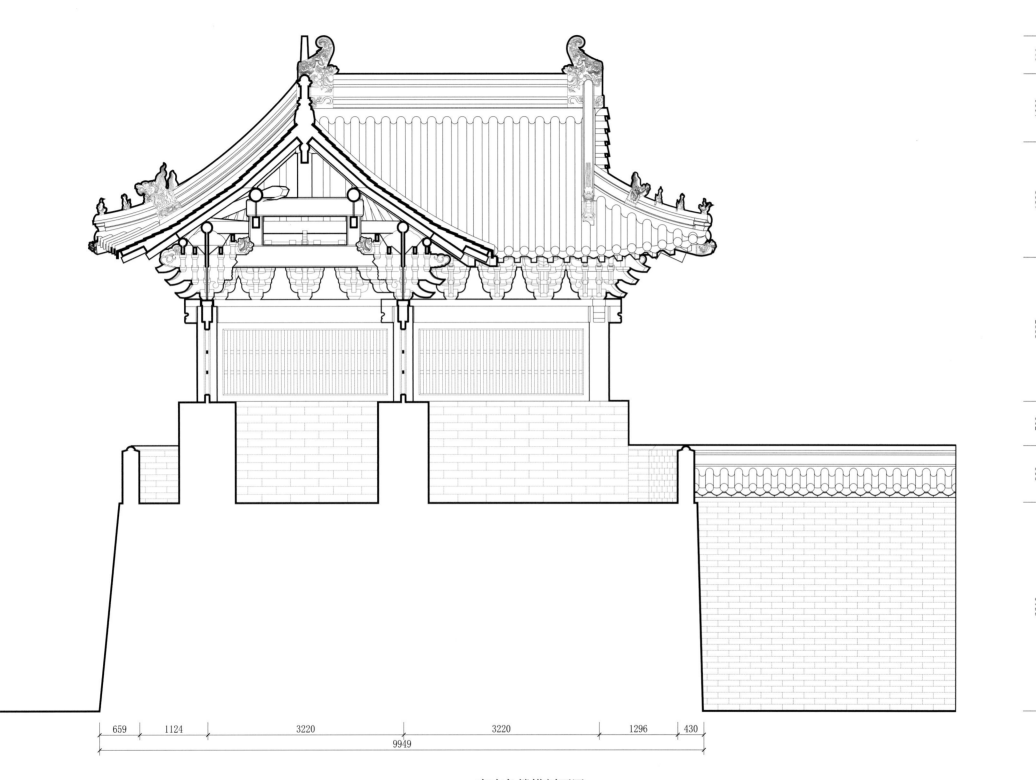

7.400
600
6.800
1080
4.410
1832
3.887
2287
1.600
700
0.900
900
±0.000
3310
-3.310

659 1124 3220 3220 1296 430
9949

东南角楼横剖面图
Cross-section of southeast corner building

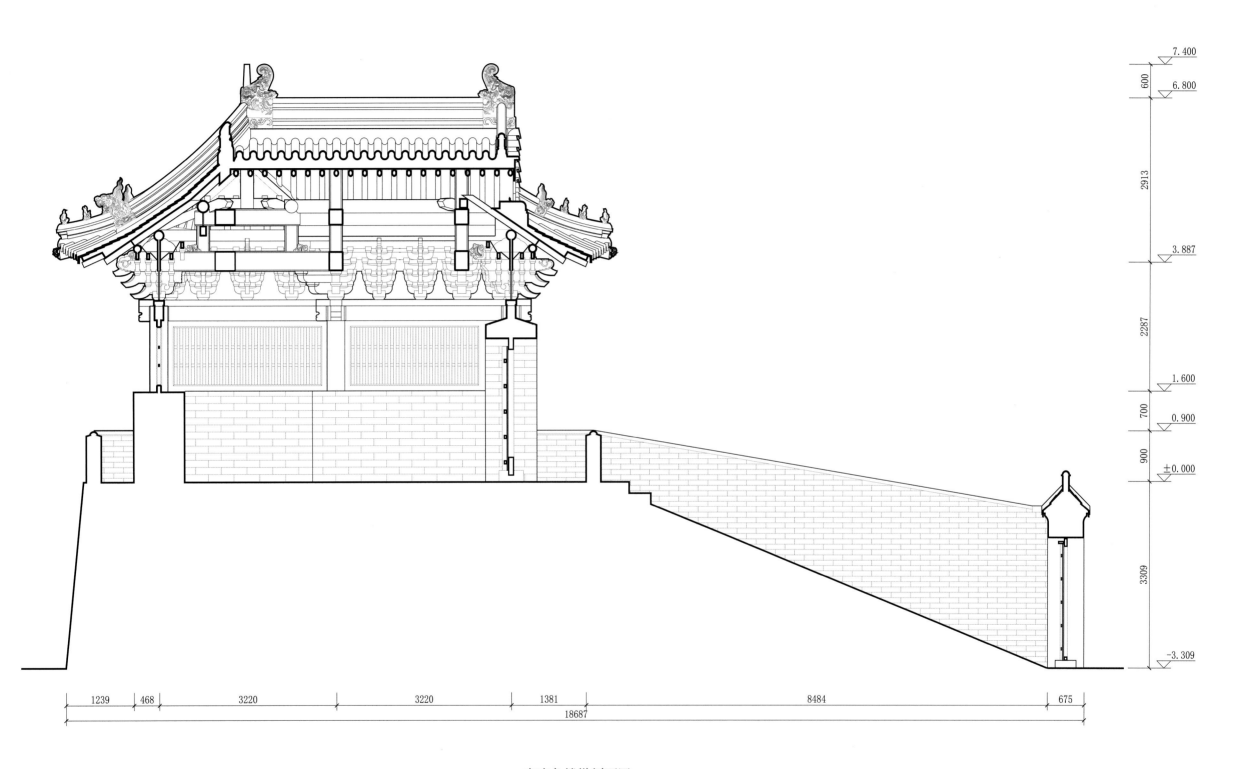

东南角楼纵剖面图
Longitudinal section of southeast corner building

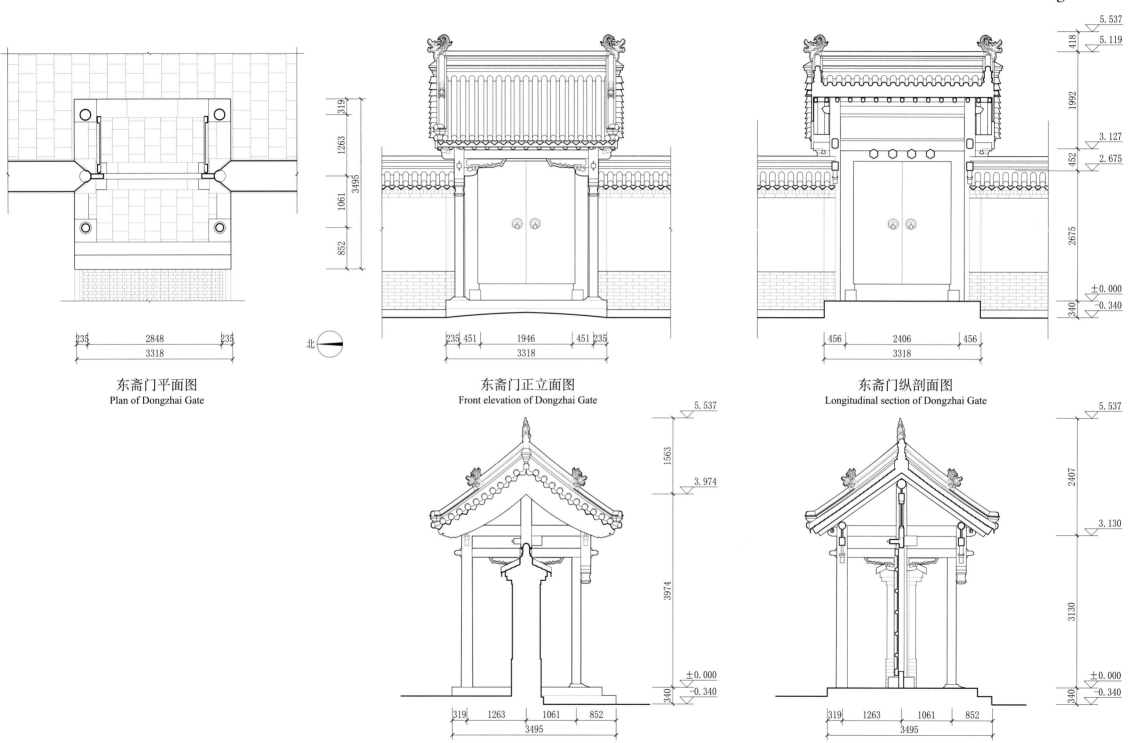

中国古建筑测绘大系·祠庙建筑——曲阜孔庙

东斋门平面图
Plan of Dongzhai Gate

东斋门正立面图
Front elevation of Dongzhai Gate

东斋门纵剖面图
Longitudinal section of Dongzhai Gate

东斋门侧立面图
Side elevation of Dongzhai Gate

东斋门横剖面图
Cross-section of Dongzhai Gate

北

中
国
古
建
筑
测
绘
大
系
·
祠
庙
建
筑
——
曲
阜
孔
庙

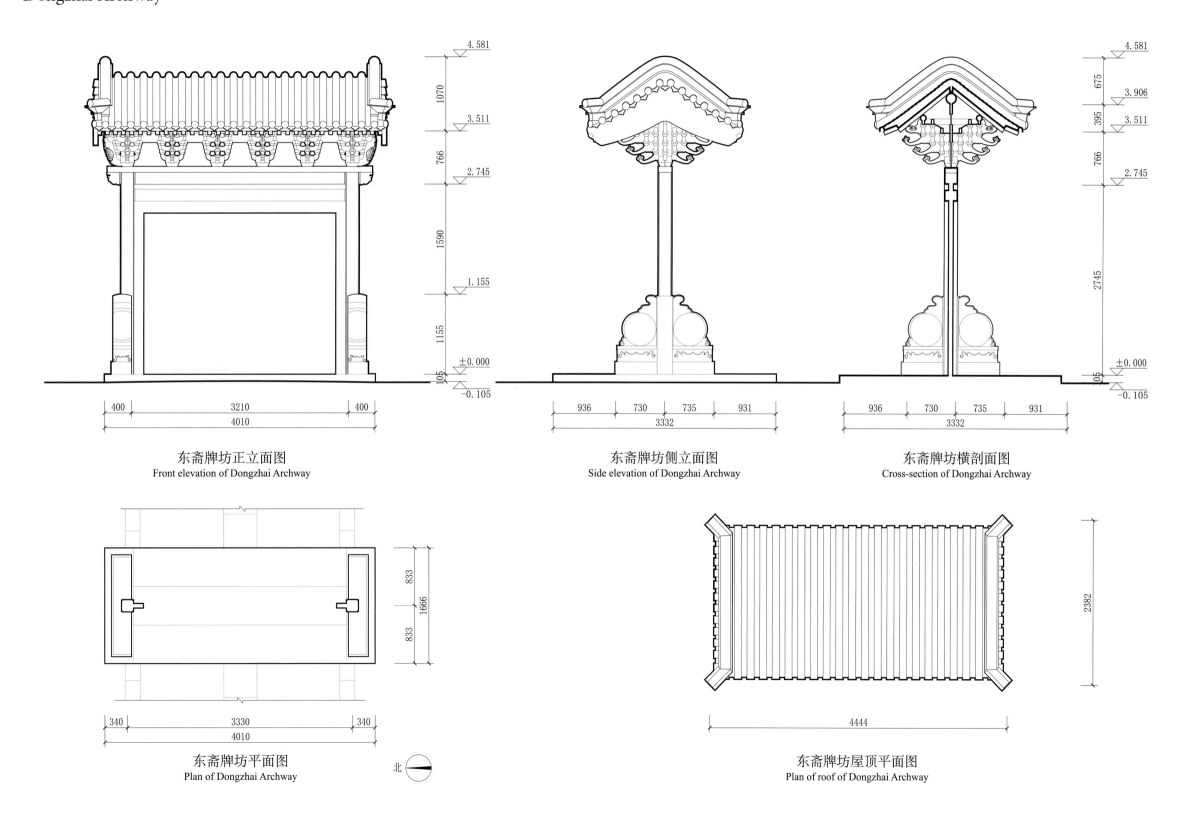

东斋牌坊正立面图
Front elevation of Dongzhai Archway

东斋牌坊侧立面图
Side elevation of Dongzhai Archway

东斋牌坊横剖面图
Cross-section of Dongzhai Archway

东斋牌坊平面图
Plan of Dongzhai Archway

北

东斋牌坊屋顶平面图
Plan of roof of Dongzhai Archway

中国古建筑测绘大系·祠庙建筑——曲阜孔庙

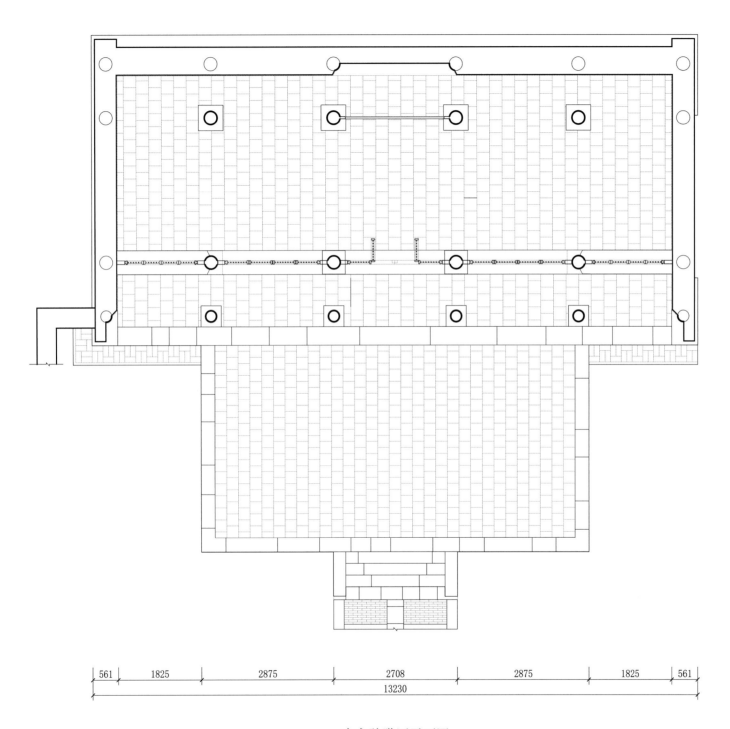

北

东斋驻跸厅平面图
Plan of Dongzhai zhubi Hall

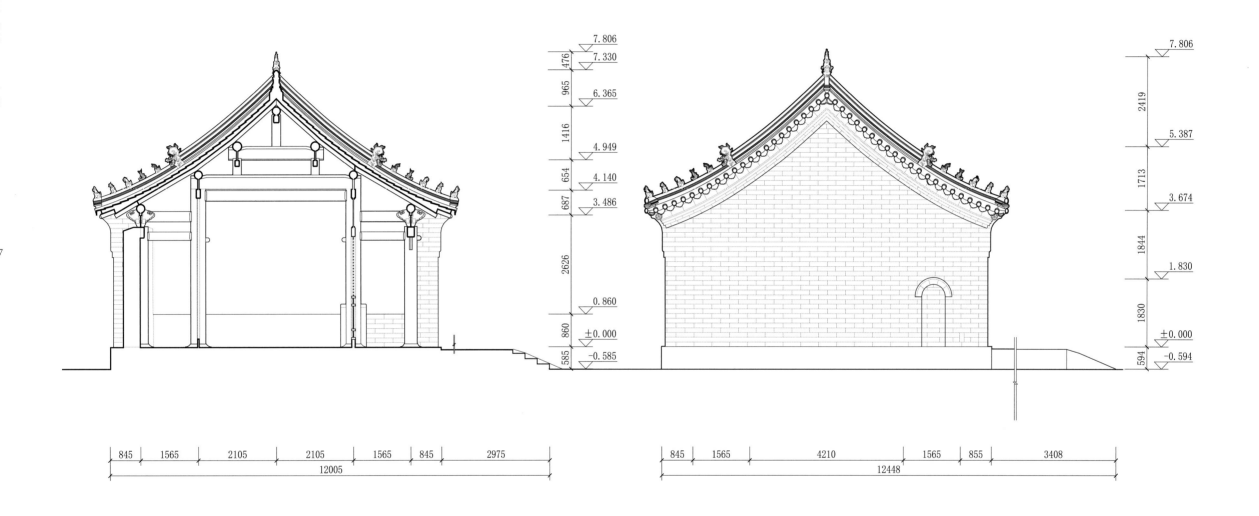

东斋驻跸厅横剖面图
Cross-section of Dongzhai zhubi Hall

东斋驻跸厅侧立面图
Side elevation of Dongzhai zhubi Hall

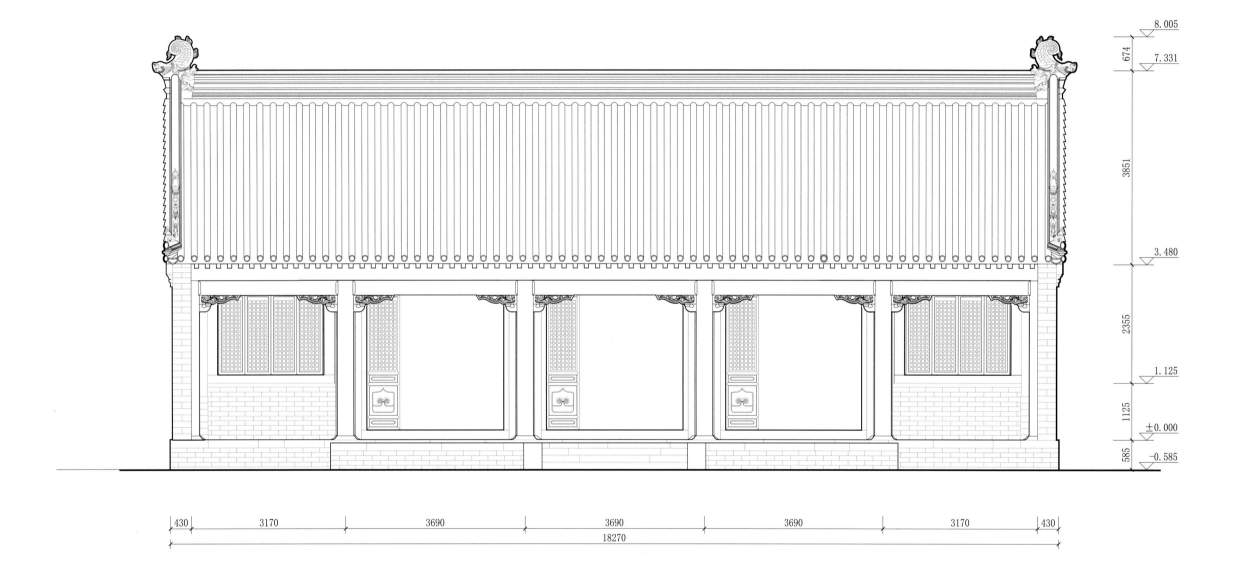

东斋驻跸厅正立面图
Front elevation of Dongzhai zhubi Hall

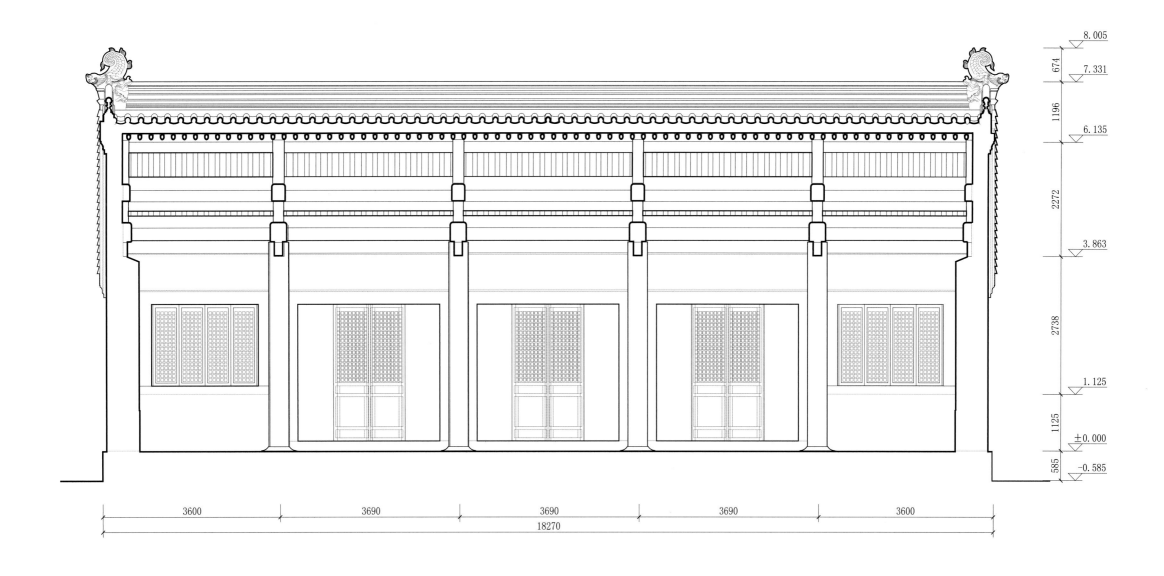

东斋驻跸厅纵剖面图
Longitudinal section of Dongzhai zhubi Hall

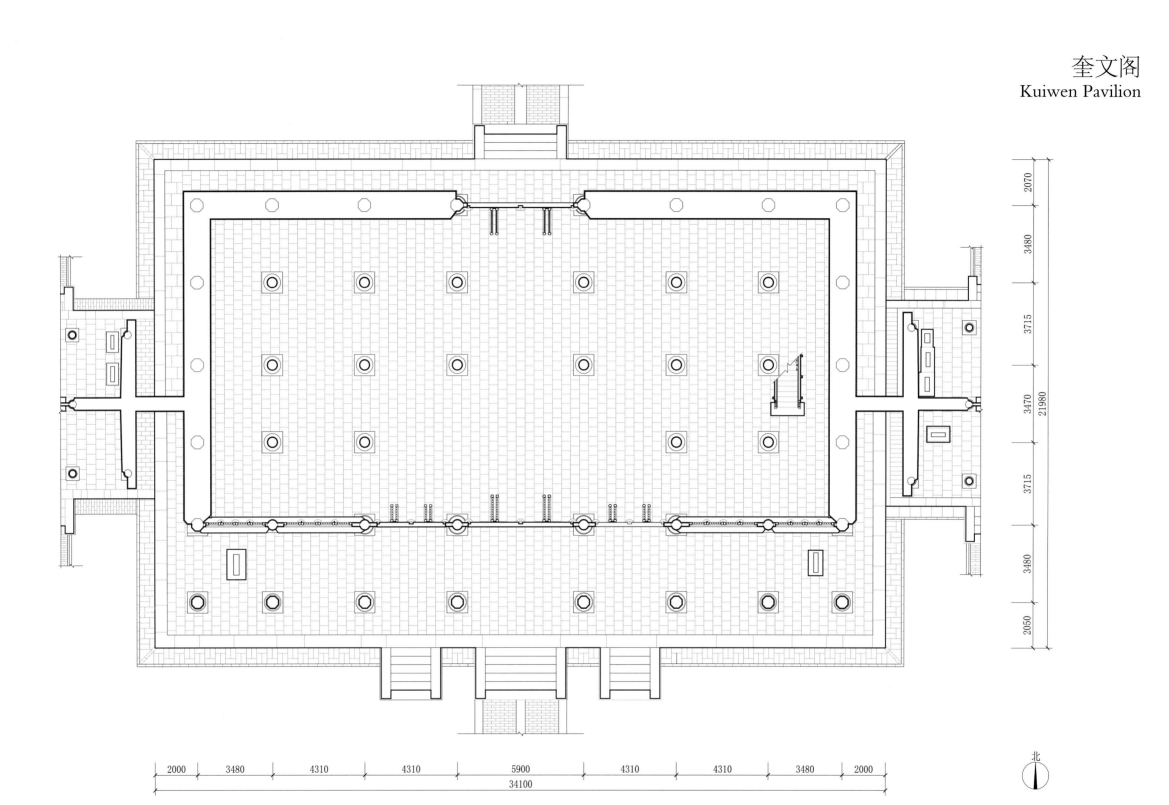

2070
3480
3715
3470
3715
3480
2050
21980

2000 3480 4310 4310 5900 4310 4310 3480 2000
34100

北

奎文阁底层平面图
Plan of ground floor of Kuiwen Pavilion

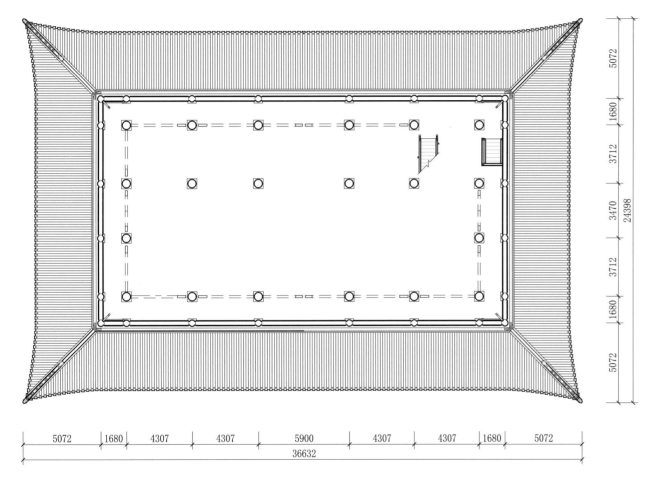

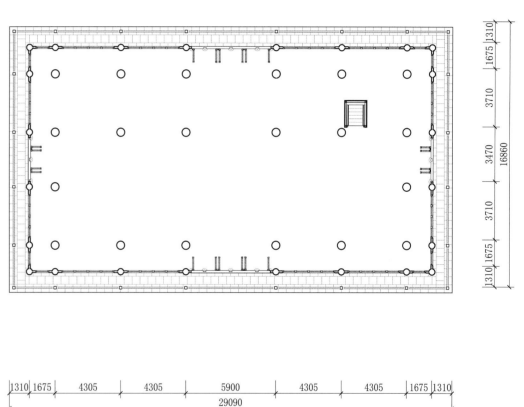

奎文阁夹层平面图
Plan of mezzanine of Kuiwen Pavilion

奎文阁二层平面图
Plan of second floor of Kuiwen Pavilion

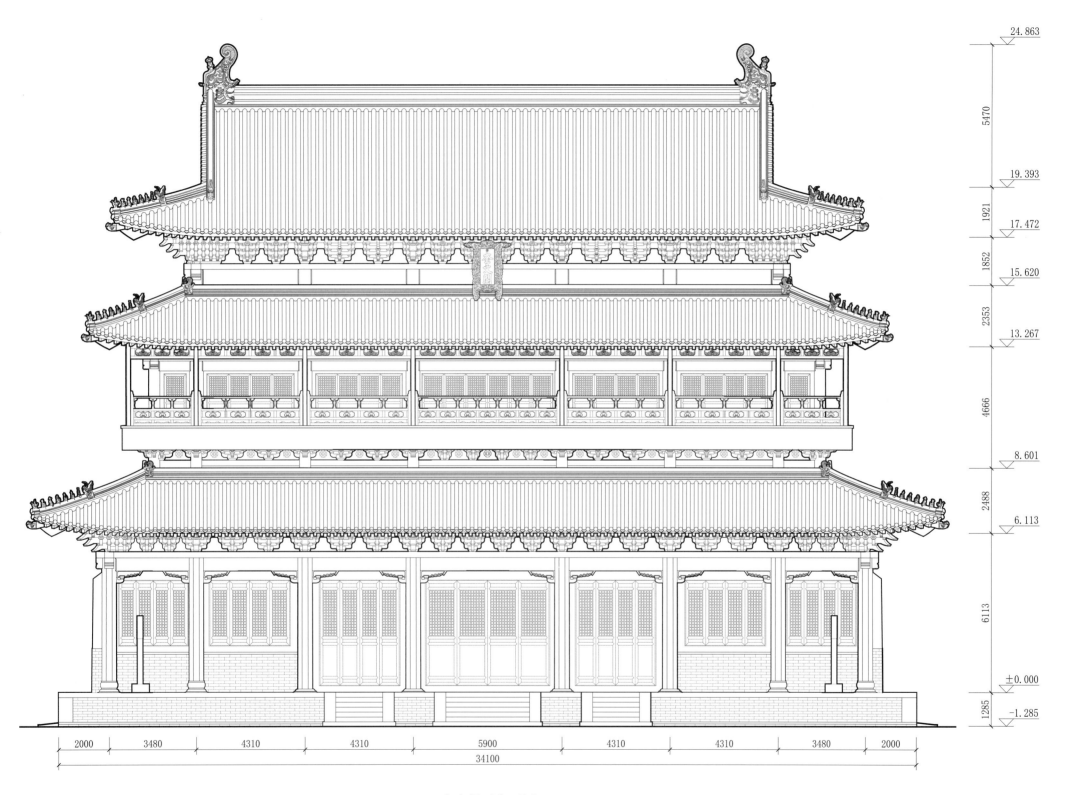

24.863

5470

19.393

1921

17.472

1852

15.620

2353

13.267

4666

8.601

2488

6.113

6113

±0.000

1285

-1.285

2000 3480 4310 4310 5900 4310 4310 3480 2000

34100

奎文阁正立面图
Front elevation of Kuiwen Pavilion

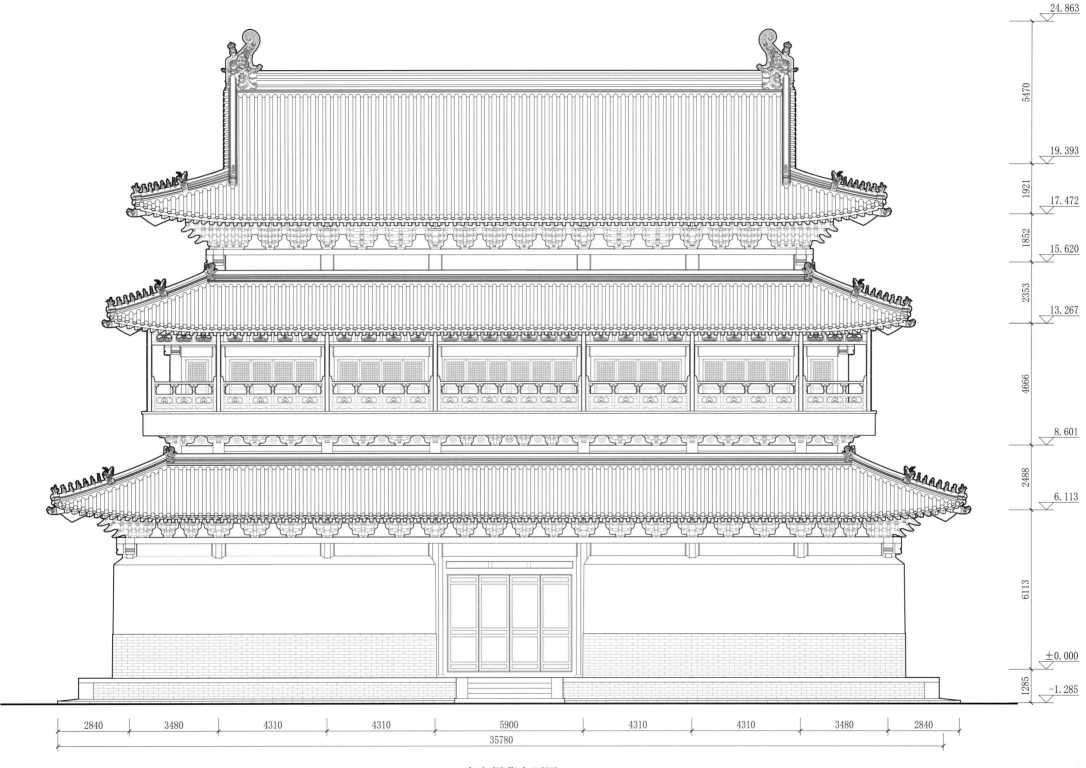

24.863

5470

19.393

1921

17.472

1852

15.620

2353

13.267

4666

8.601

2488

6.113

6113

±0.000

1285

-1.285

2840　3480　4310　4310　5900　4310　4310　3480　2840

35780

奎文阁背立面图
Rear elevation of Kuiwen Pavilion

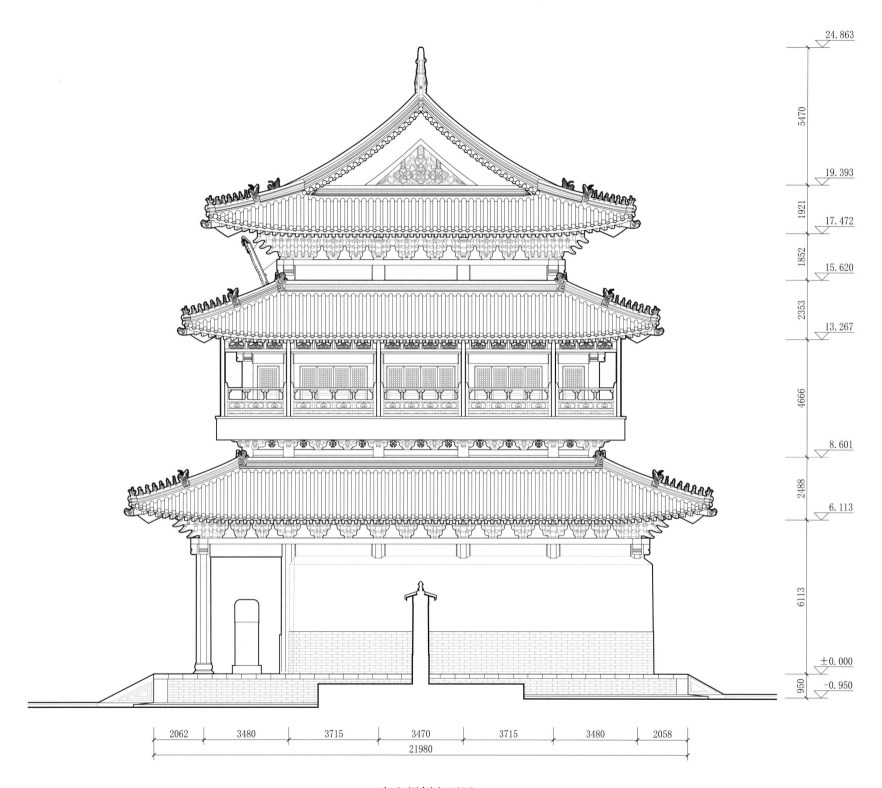

24.863

5470

19.393

1921

17.472

1852

15.620

2353

13.267

4666

8.601

2488

6.113

6113

±0.000

950

-0.950

2062 3480 3715 3470 3715 3480 2058

21980

奎文阁侧立面图

Side elevation of Kuiwen Pavilion

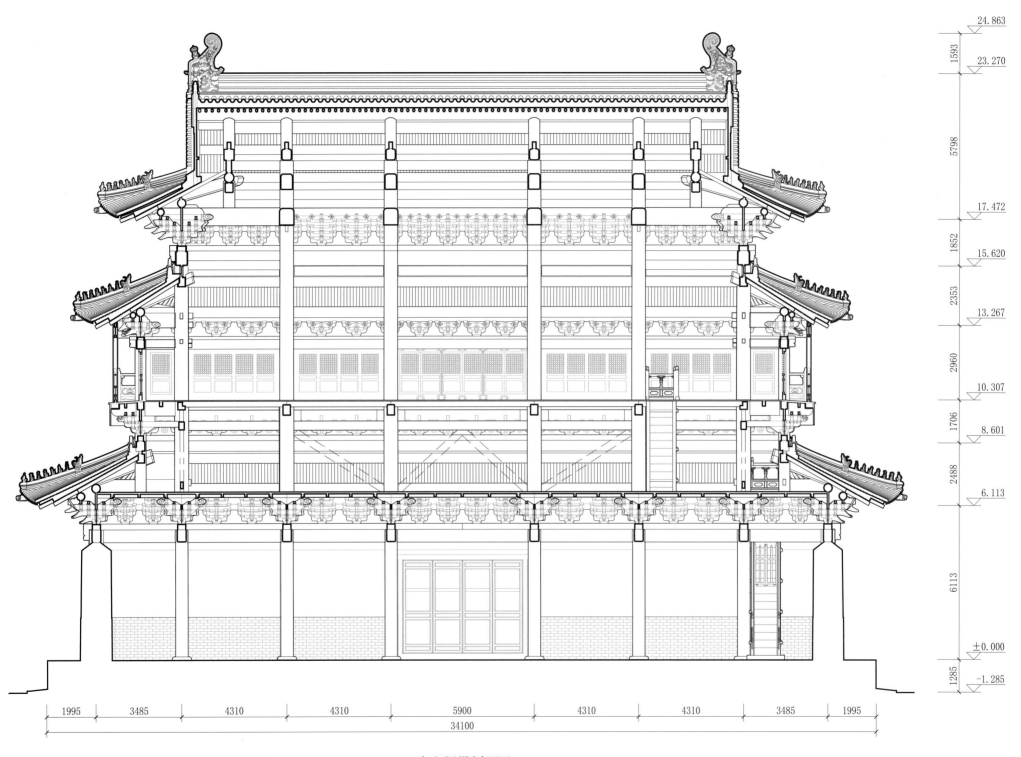

24.863
1593
23.270
5798
17.472
1852
15.620
2353
13.267
2960
10.307
1706
8.601
2488
6.113
6113
±0.000
1285
-1.285

1995 3485 4310 4310 5900 4310 4310 3485 1995
34100

奎文阁纵剖面图
Longitudinal section of Kuiwen Pavilion

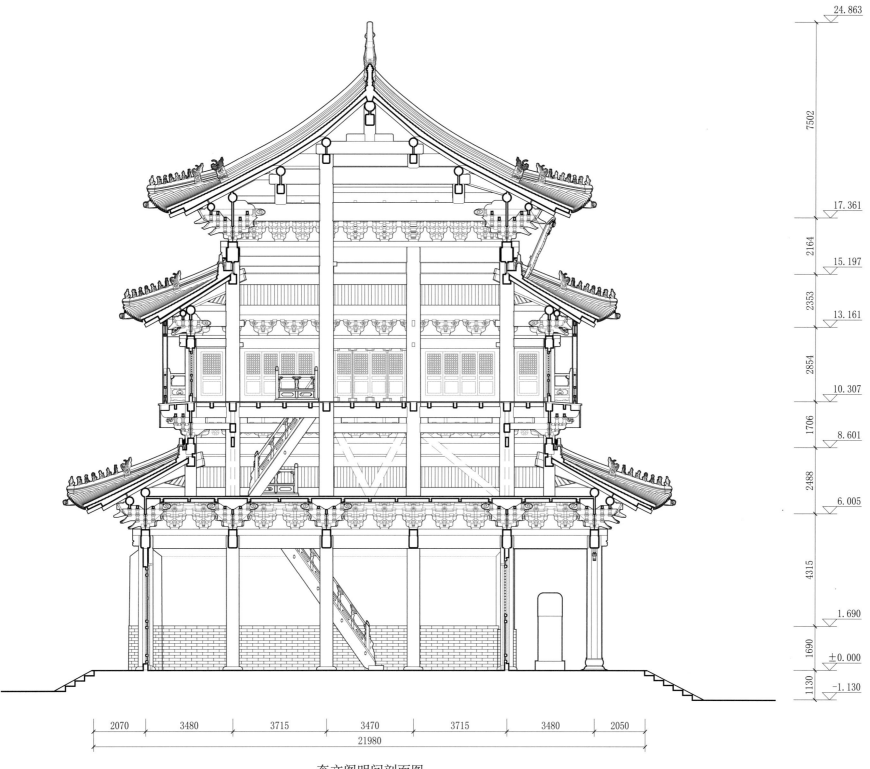

24.863

7502

17.361

2164

15.197

2353

13.161

2854

10.307

1706

8.601

2488

6.005

4315

1.690

1690

±0.000

1130

-1.130

2070　3480　3715　3470　3715　3480　2050

21980

奎文阁明间剖面图

Section of central bay of Kuiwen Pavilion

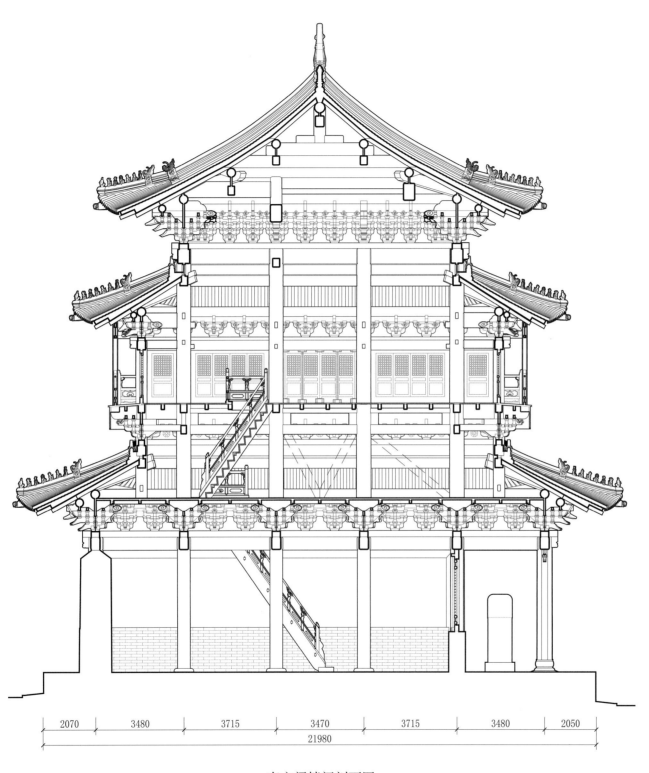

24.863

1593

23.270

5908

17.362

1745

15.616

2454

13.162

2855

10.307

1614

8.693

2688

6.005

4315

1.690

1690

±0.000

1285

-1.285

2070　3480　3715　3470　3715　3480　2050

21980

奎文阁梢间剖面图

Section of second-to-last bay of Kuiwen Pavilion

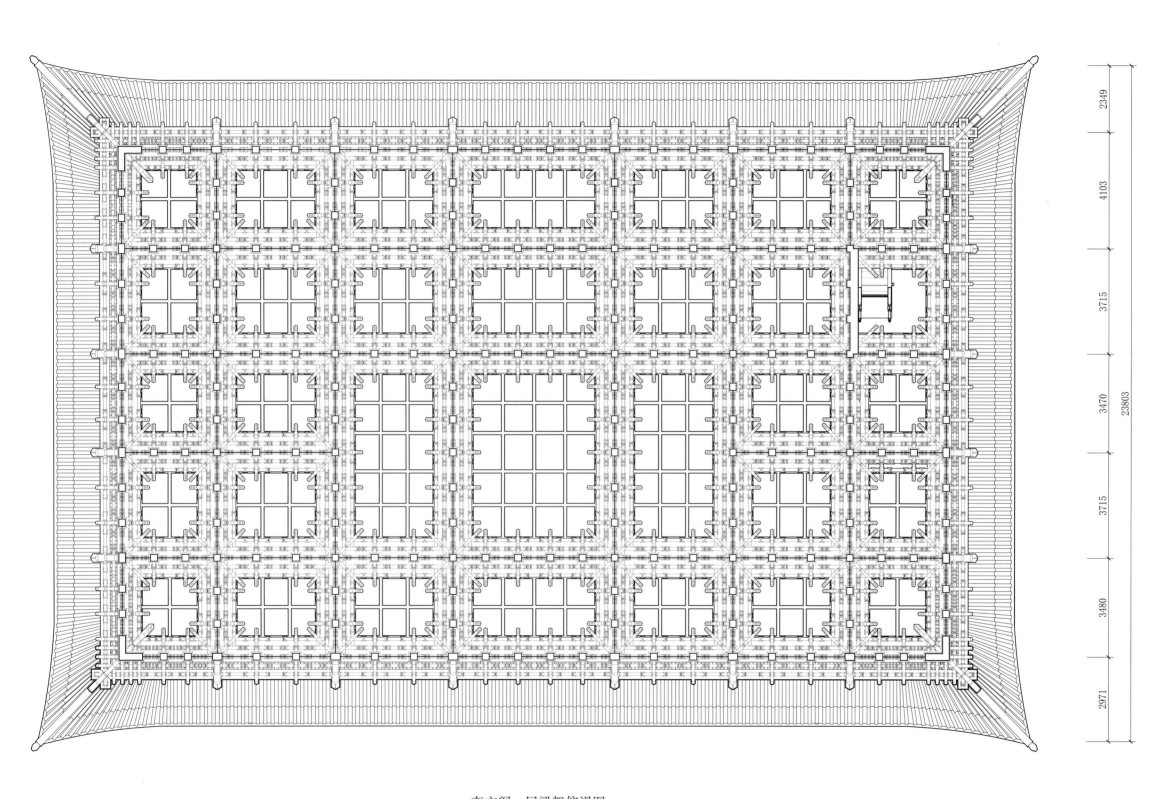

奎文阁一层梁架仰视图

Plan of first-floor framework of Kuiwen Pavilion as seen from below

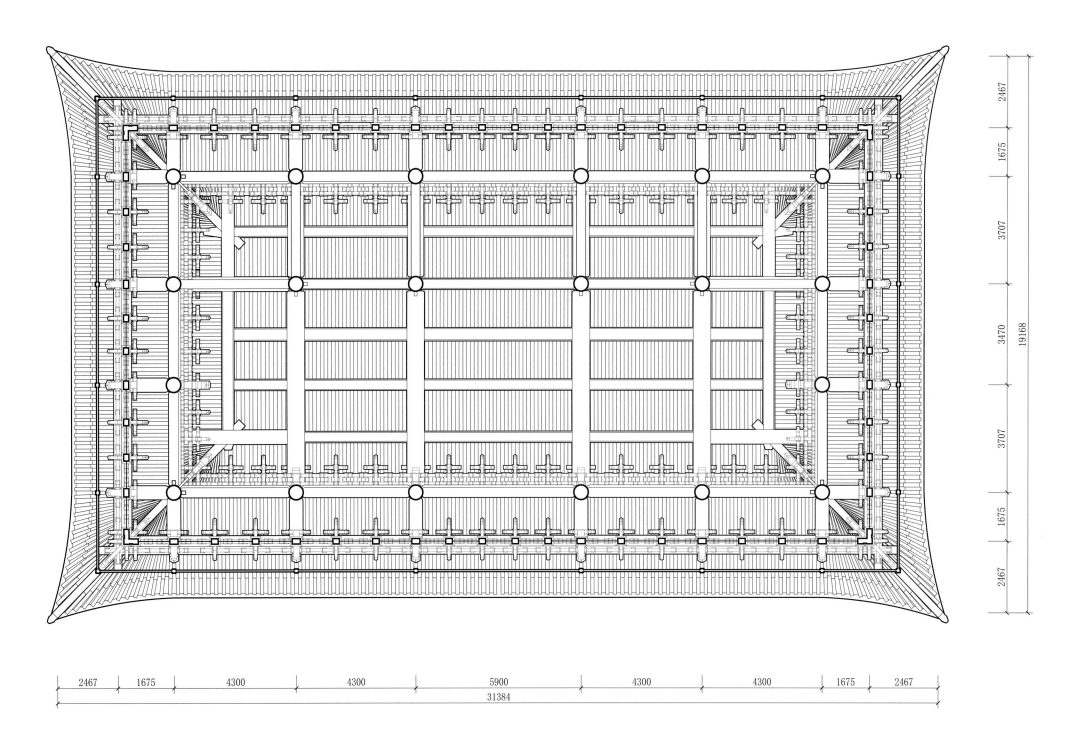

奎文阁二层下檐梁架仰视图
Plan of second-floor lower eaves framework of Kuiwen Pavilion as seen from below

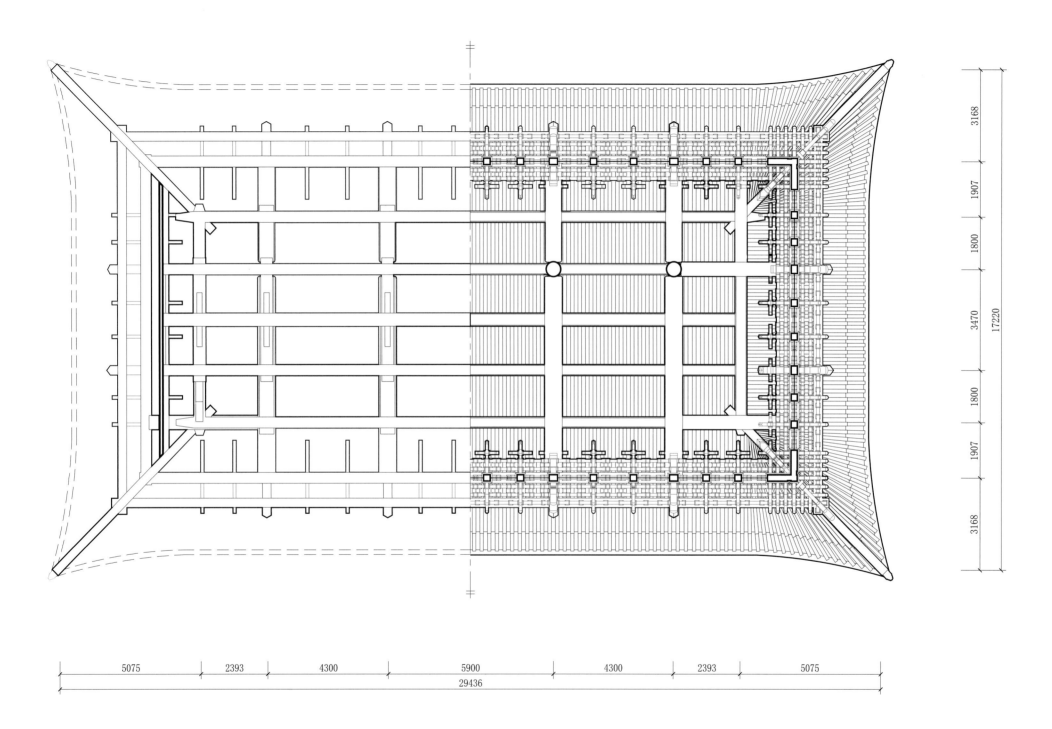

奎文阁二层上檐梁架仰俯视图

Plan of second-floor upper eaves framework of Kuiwen Pavilion as seen from below

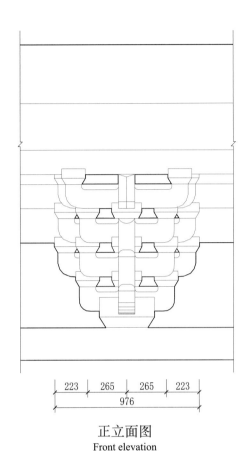

正立面图
Front elevation

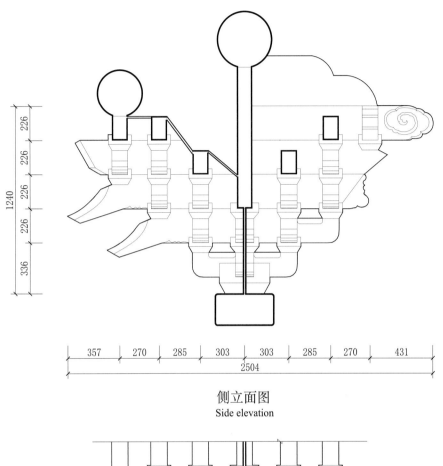

侧立面图
Side elevation

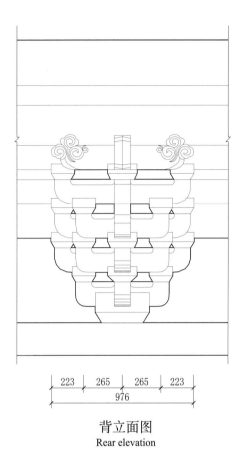

背立面图
Rear elevation

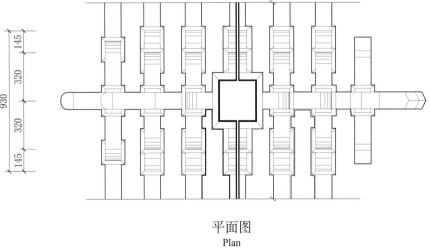

平面图
Plan

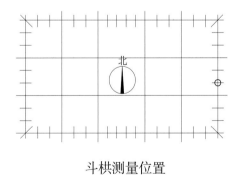

斗栱测量位置

奎文阁二层上檐平身科斗栱
Intercolumnar bracket sets of second-floor upper eaves of Kuiwen Pavilion

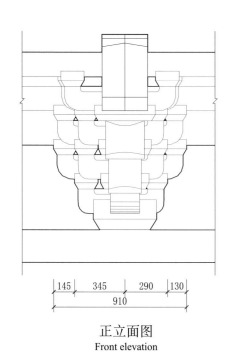

正立面图
Front elevation

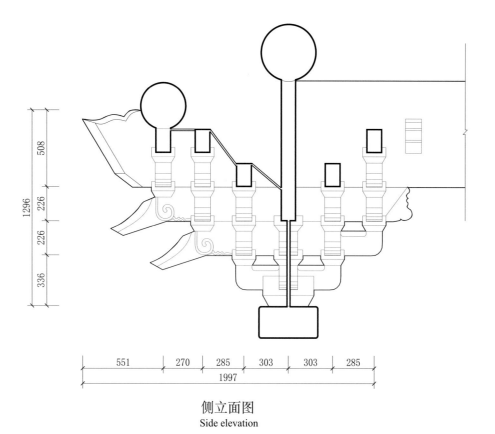

侧立面图
Side elevation

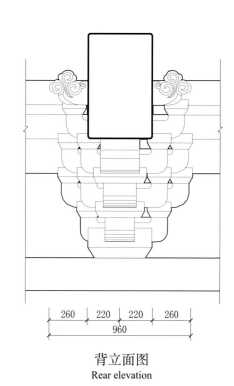

背立面图
Rear elevation

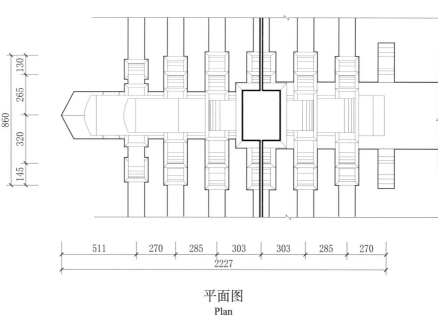

平面图
Plan

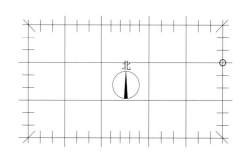

斗栱测量位置

奎文阁二层上檐柱头科斗栱
Bracket sets atop columns of second-floor upper eaves of Kuiwen Pavilion

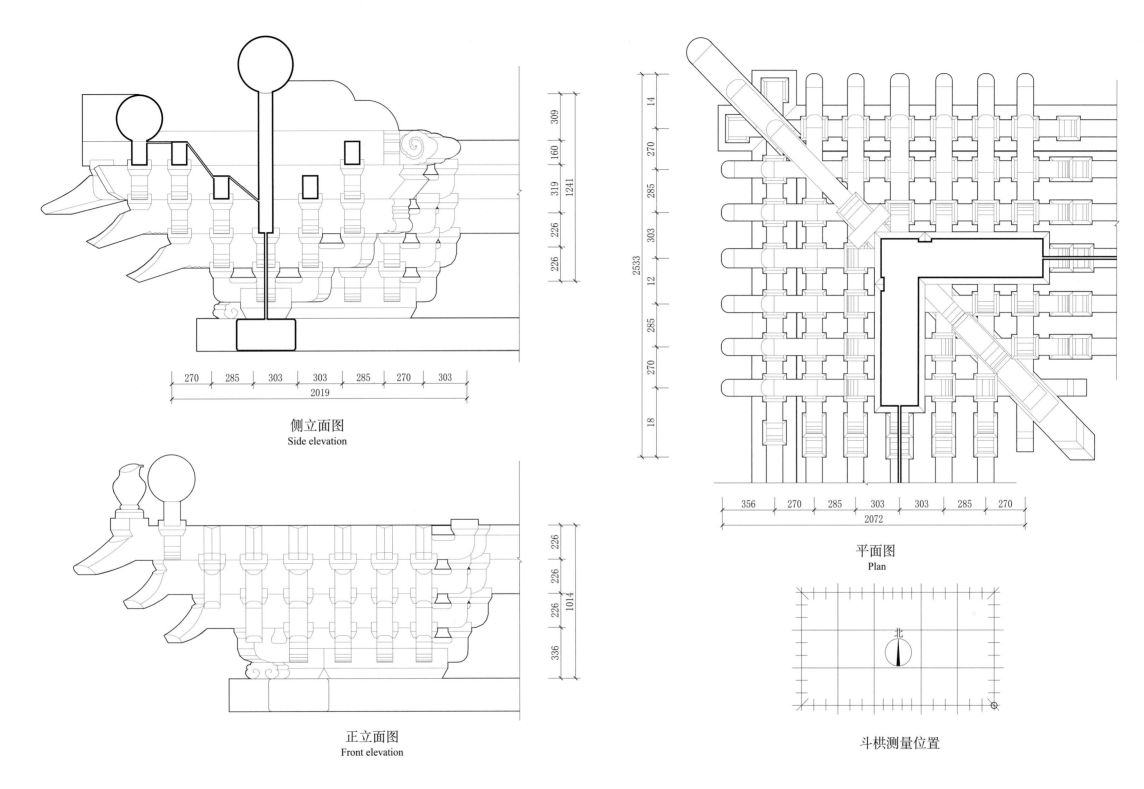

侧立面图
Side elevation

正立面图
Front elevation

平面图
Plan

北

斗栱测量位置

奎文阁二层上檐斗栱角科
Corner bracket sets of second-floor upper eaves of Kuiwen Pavilion

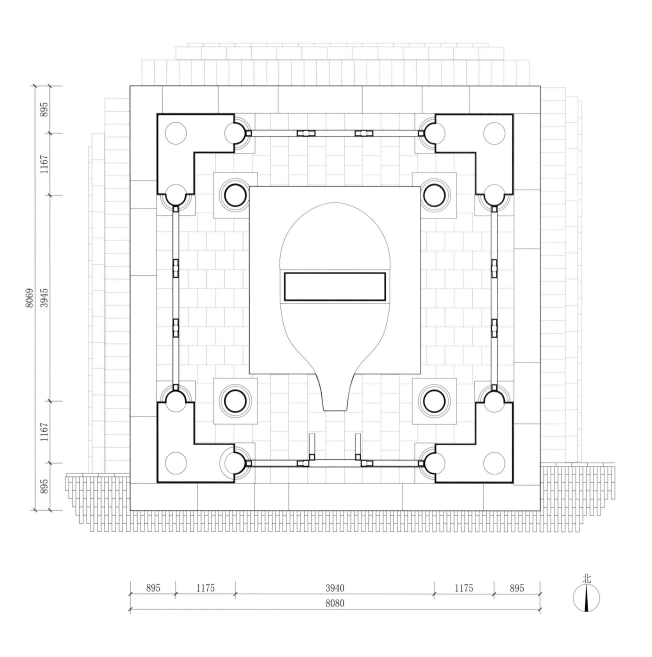

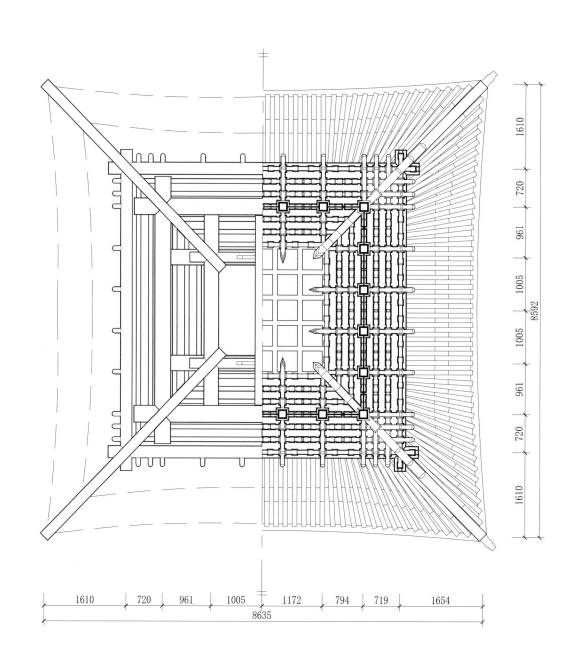

北

清雍正八年碑西碑亭（一号碑亭）平面图

Plan of west stele pavilion of the eighth reign year of emperor Yongzheng,Qing Dynasty (No.1 stele pavilion)

清雍正八年碑西碑亭（一号碑亭）上檐梁架仰俯视图

Plan of upper eaves framework of west tele pavilion of the eighth reign year of emperor Yongzheng,Qing Dynasty,(No.1 stele pavilion), as seen from below

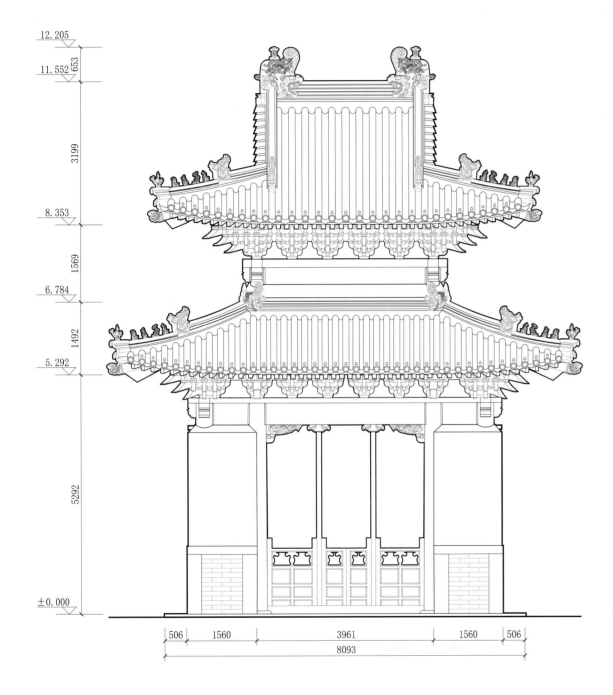

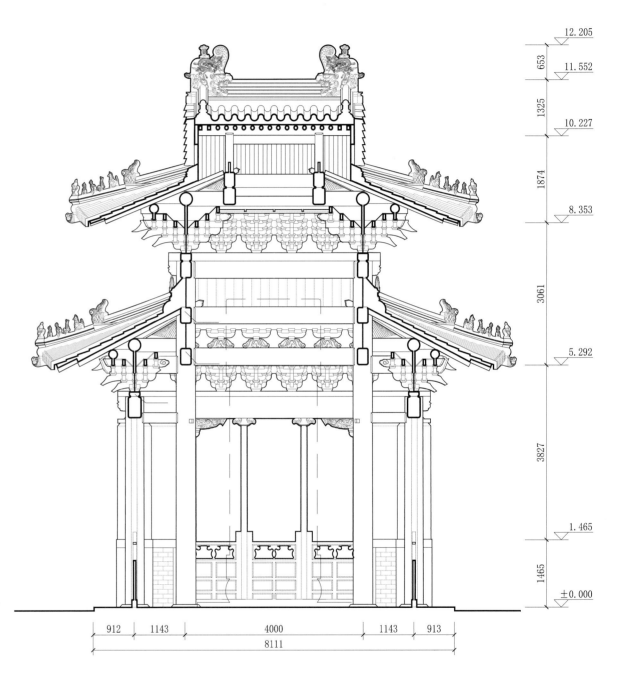

清雍正八年碑西碑亭（一号碑亭）正立面图
Front elevation of west stele pavilion of the eighth reign year of emperor Yongzheng,Qing Dynasty (No.1 stele pavilion)

清雍正八年碑西碑亭（一号碑亭）纵剖面图
Longitudinal section of west stele pavilion of the eighth reign year of emperor Yongzheng,Qing Dynasty (No.1 stele pavilion)

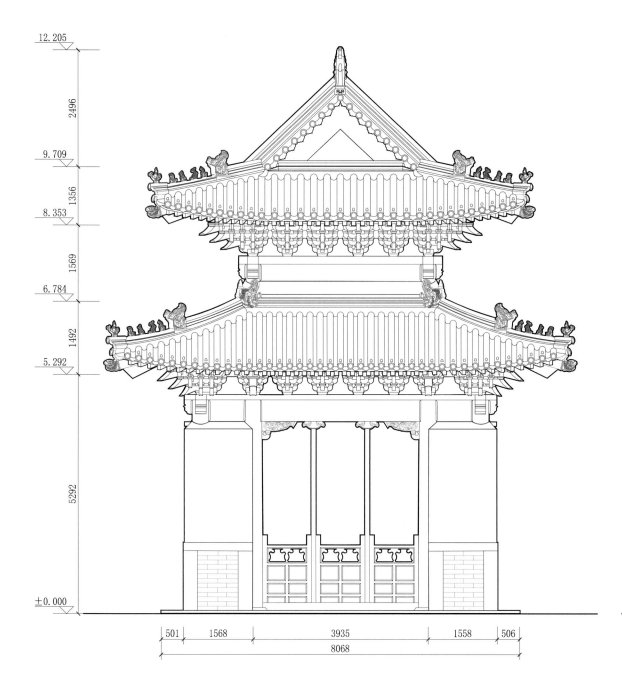

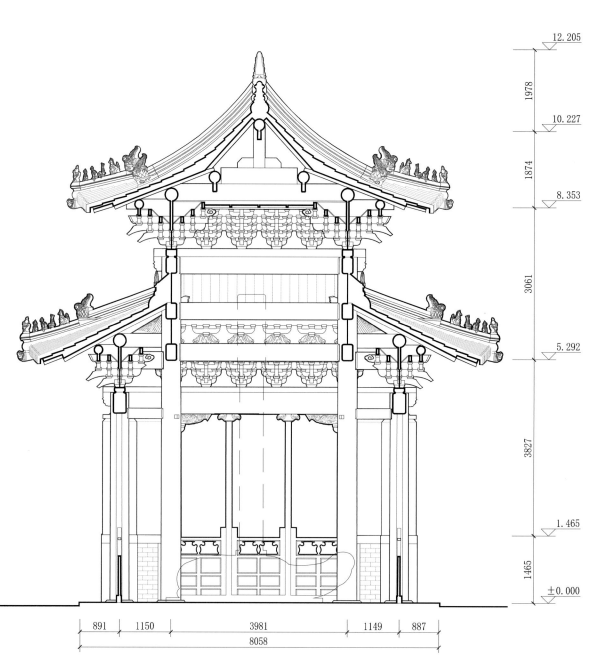

清雍正八年碑西碑亭（一号碑亭）侧立面图
Side elevation of west stele pavilion of the eighth reign year of emperor Yongzheng,Qing Dynasty (No.1 stele pavilion)

清雍正八年碑西碑亭（一号碑亭）横剖面图
Cross-section of west stele pavilion of the eighth reign year of emperor Yongzheng,Qing Dynasty (No.1 stele pavilion)

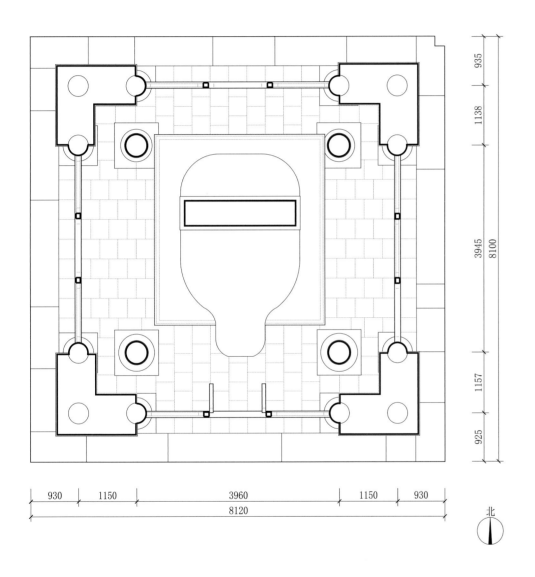

清康熙三十二年碑碑亭（二号碑亭）
Stele pavilion of the thirty-second reign year of emperor Kangxi,Qing Dynasty (No.2 stele pavilion)

中国古建筑测绘大系·祠庙建筑——曲阜孔庙

097

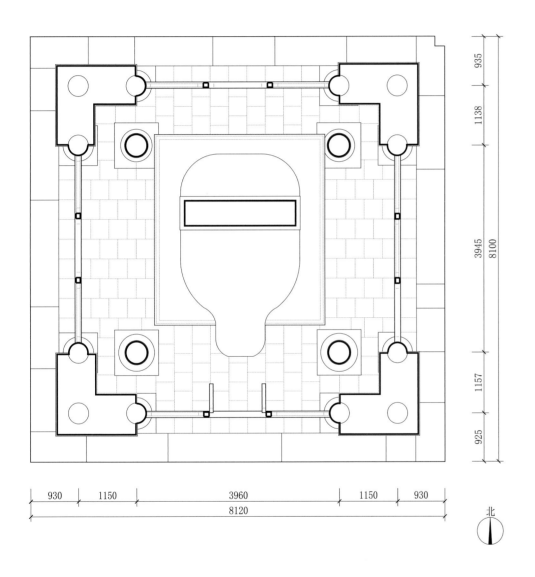

北

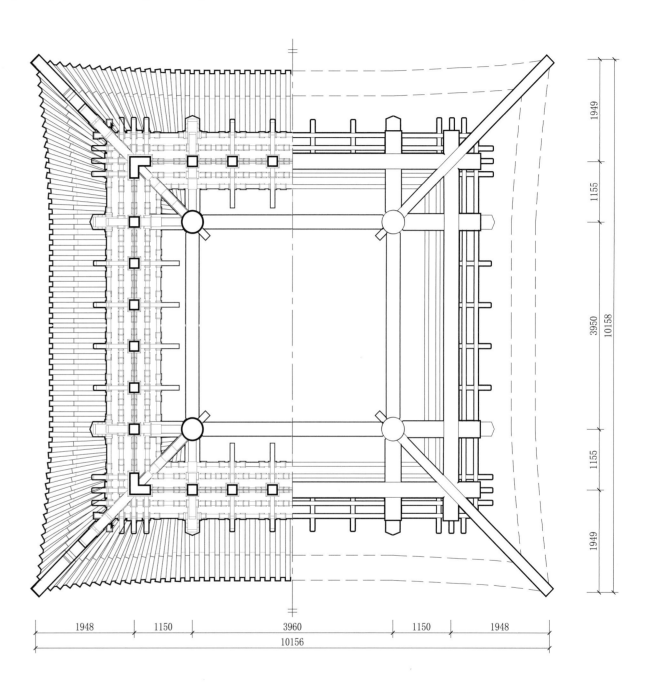

清康熙三十二年碑碑亭（二号碑亭）平面图
Plan of stele pavilion of the thirty-second reign year of emperor Kangxi,Qing Dynasty (No.2 stele pavilion)

清康熙三十二年碑碑亭（二号碑亭）下檐梁架仰俯视图
Plan of lower eaves framework of stele pavilion of the thirty-second reign year of emperor Kangxi,Qing Dynasty (No.2 stele pavilion), as seen from below

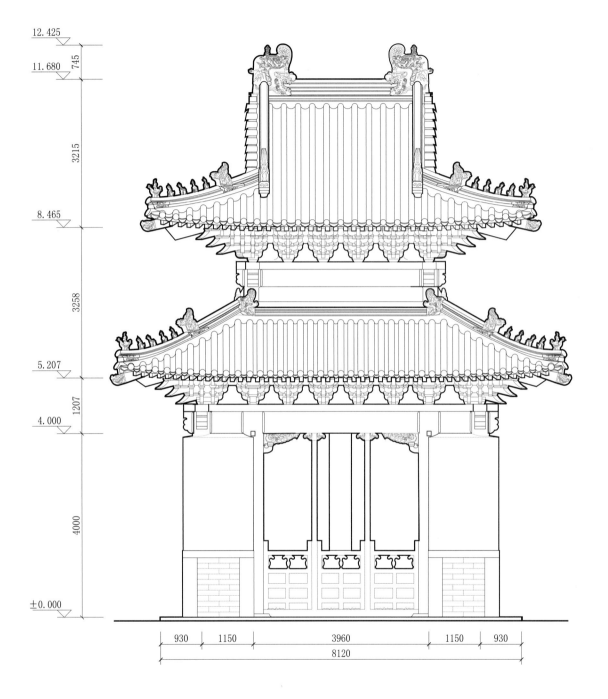

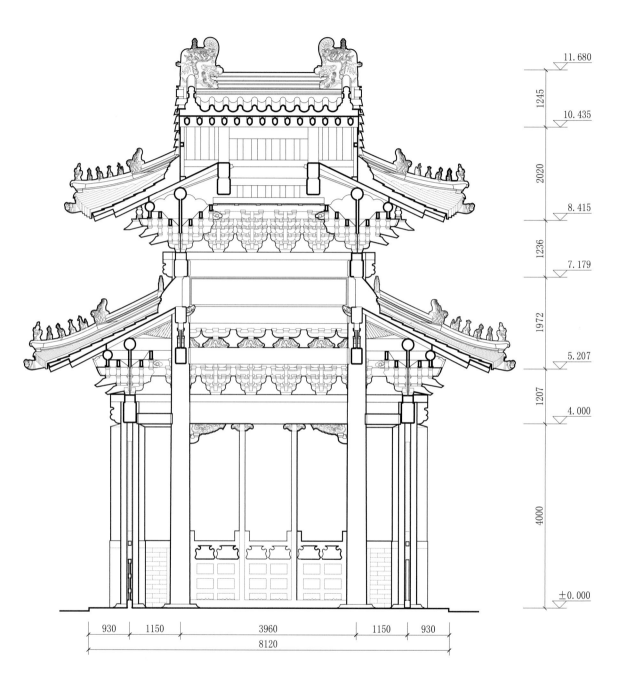

清康熙三十二年碑碑亭（二号碑亭）正立面图
Front elevation of stele pavilion of the thirty-second reign year of emperor Kangxi,Qing Dynasty (No.2 stele pavilion)

清康熙三十二年碑碑亭（二号碑亭）纵剖面图
Longitudinal section of stele pavilion of the thirty-second reign year of emperor Kangxi,Qing Dynasty (No.2 stele pavilion)

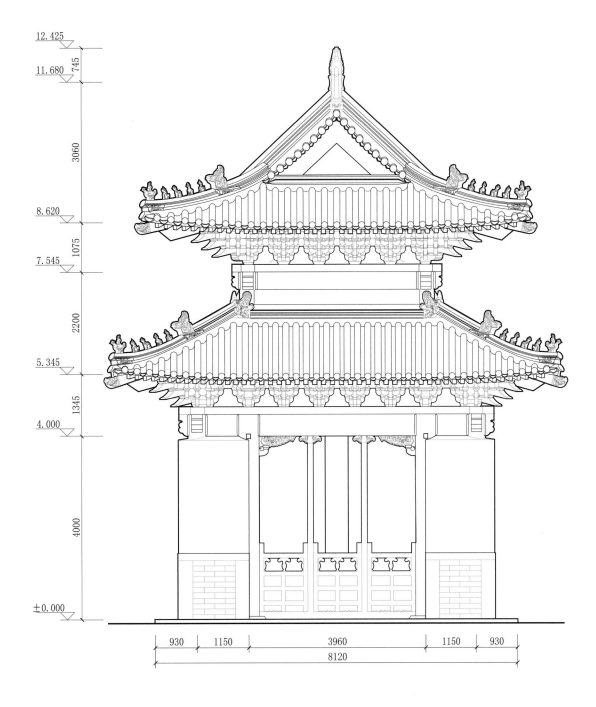

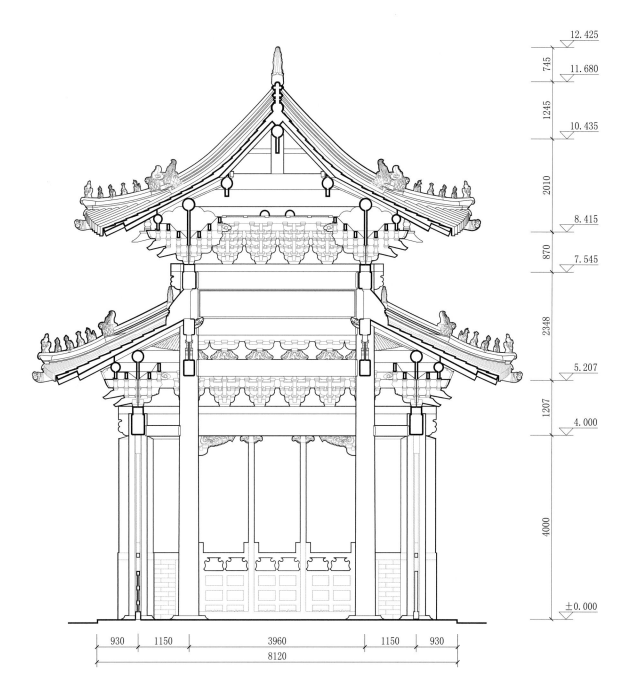

清康熙三十二年碑碑亭（二号碑亭）侧立面图

Side elevation of stele pavilion of the thirty-second reign year of emperor Kangxi,Qing Dynasty (No.2 stele pavilion)

清康熙三十二年碑碑亭（二号碑亭）横剖面图

Cross-section of stele pavilion of the thirty-second reign year of emperor Kangxi,Qing Dynasty (No.2 stele pavilion)

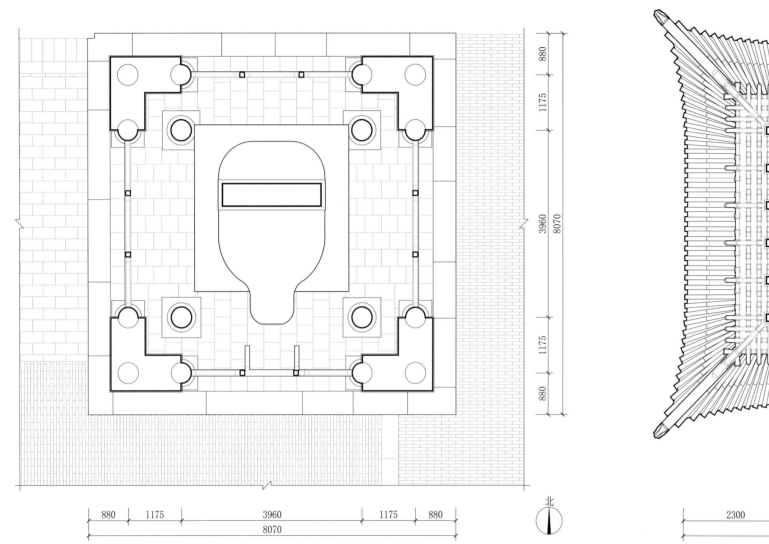

清康熙二十五年碑碑亭（三号碑亭）平面图
Plan of stele pavilion of the twenty-fifth reign year of emperor Kangxi,Qing Dynasty (No.3 stele pavilion)

清康熙二十五年碑碑亭（三号碑亭）上檐梁架仰俯视图
Plan of upper eaves framework of stele pavilion of the twenty-fifth reign year of emperor Kangxi,Qing Dynasty (No.3 stele pavilion), as seen from below

北

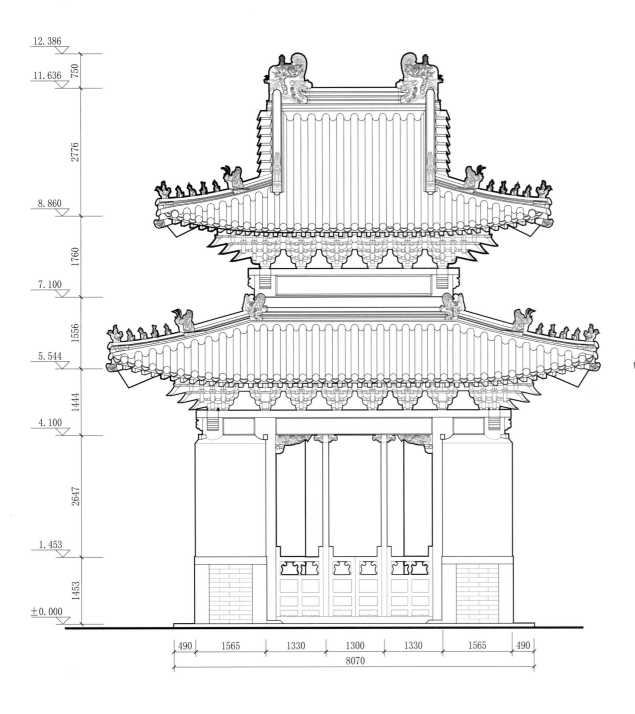

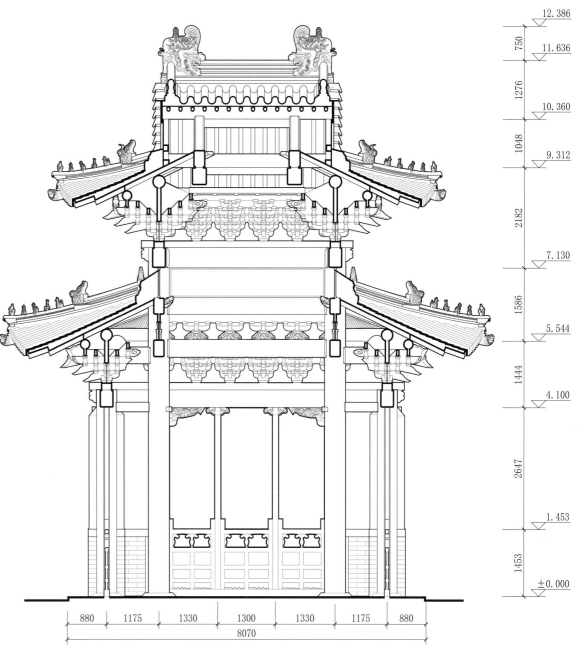

清康熙二十五年碑碑亭（三号碑亭）正立面图

Front elevation of stele pavilion of the twenty-fifth reign year of emperor Kangxi,Qing Dynasty (No.3 stele pavilion)

清康熙二十五年碑碑亭（三号碑亭）纵剖面图

Longitudinal section of stele pavilion of the twenty-fifth reign year of emperor Kangxi,Qing Dynasty (No.3 stele pavilion)

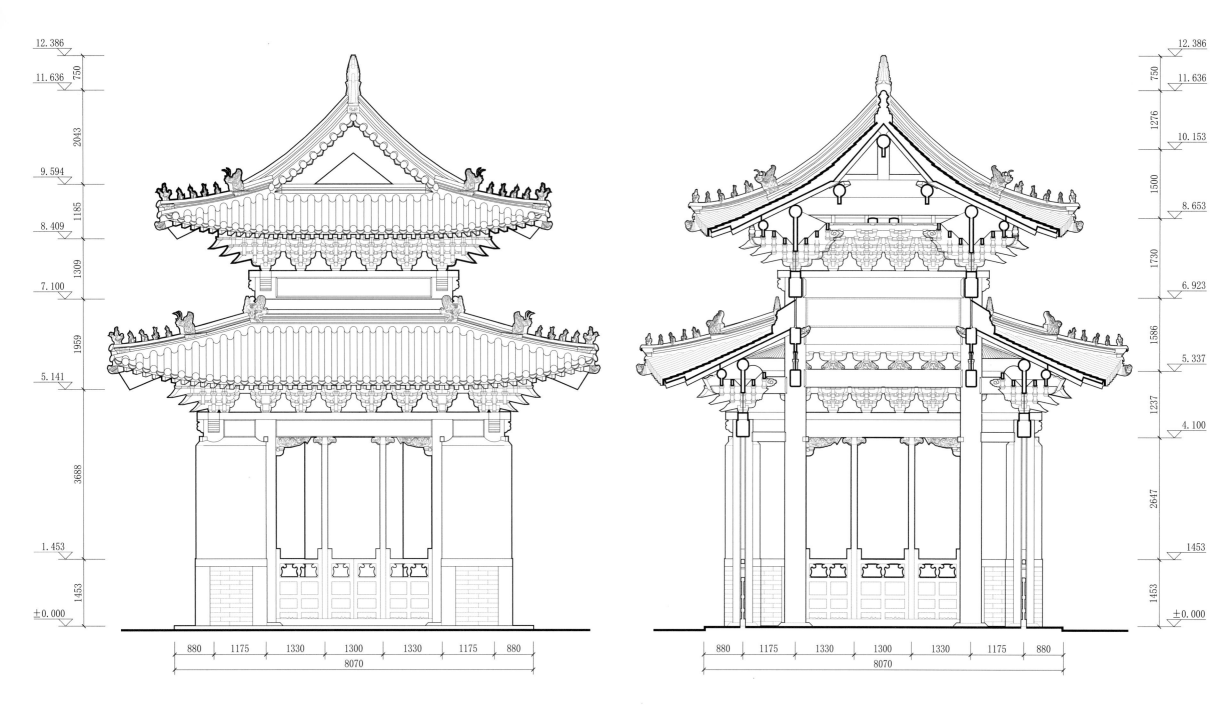

清康熙二十五年碑碑亭（三号碑亭）侧立面图
Side elevation of stele pavilion of the twenty-fifth reign year of emperor Kangxi,Qing Dynasty (No.3 stele pavilion)

清康熙二十五年碑碑亭（三号碑亭）横剖面图
Cross-section of stele pavilion of the twenty-fifth reign year of emperor Kangxi,Qing Dynasty (No.3 stele pavilion)

清雍正八年碑东碑亭（四号碑亭）

East stele pavilion of the eighth reign year of emperor Yongzheng,Qing Dynasty (No.4 stele pavilion)

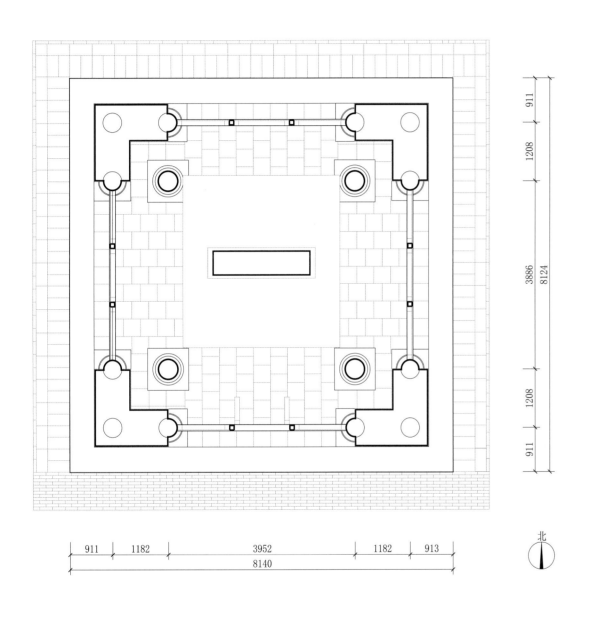

北

清雍正八年碑东碑亭（四号碑亭）平面图

Plan of east stele pavilion of the eighth reign year of emperor Yongzheng,Qing Dynasty (No.4 stele pavilion)

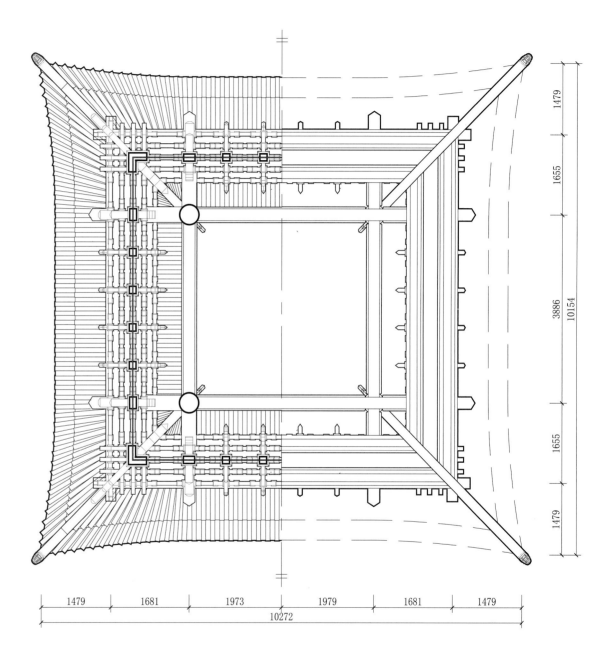

清雍正八年碑东碑亭（四号碑亭）下檐梁架仰俯视图

Plan of lower eaves framework of east tele pavilion of the eighth reign year of emperor Yongzheng,Qing Dynasty,(No.4 stele pavilion), as seen from below

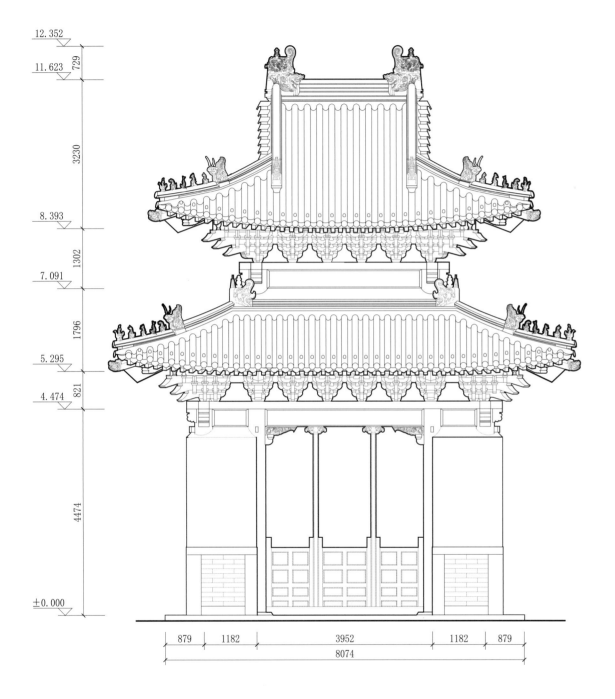

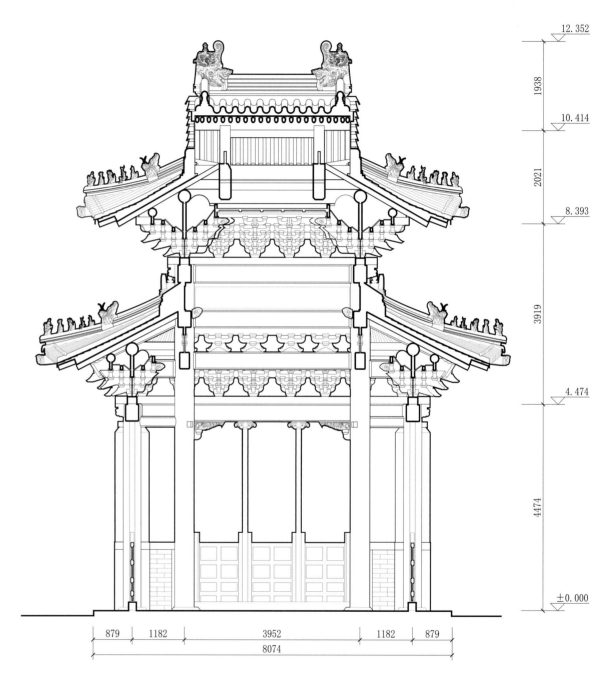

清雍正八年碑东碑亭（四号碑亭）正立面图
Front elevation of east stele pavilion of the eighth reign year of emperor Yongzheng,Qing Dynasty (No.4 stele pavilion)

清雍正八年碑东碑亭（四号碑亭）纵剖面图
Longitudinal section of east stele pavilion of the eighth reign year of emperor Yongzheng,Qing Dynasty (No.4 stele pavilion)

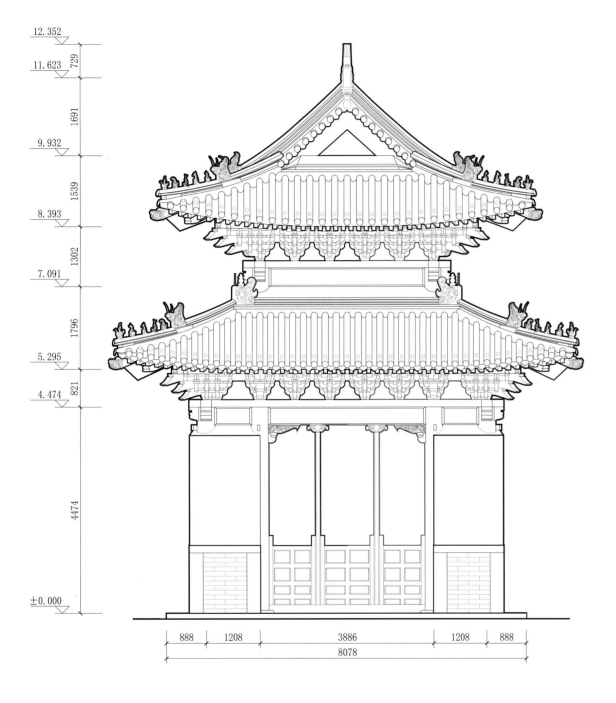

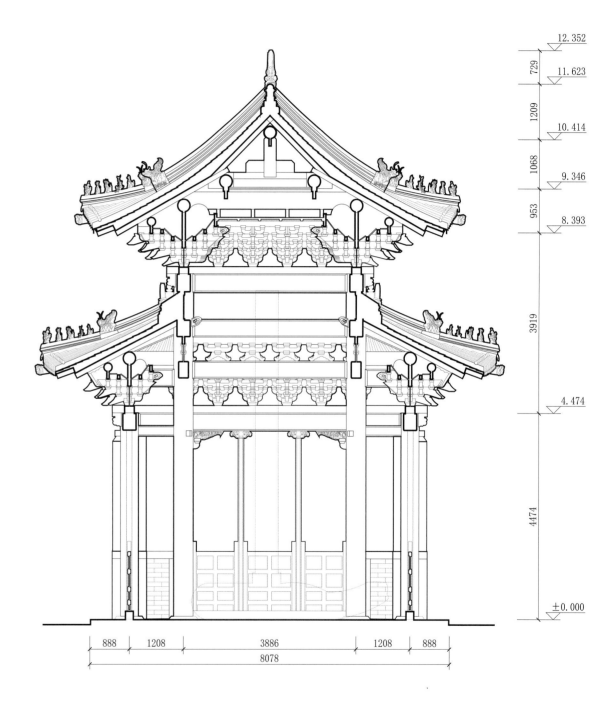

清雍正八年碑东碑亭（四号碑亭）侧立面图
Side elevation of east stele pavilion of the eighth reign year of emperor Yongzheng,Qing Dynasty (No.4 stele pavilion)

清雍正八年碑东碑亭（四号碑亭）横剖面图
Cross-section of east stele pavilion of the eighth reign year of emperor Yongzheng,Qing Dynasty (No.4 stele pavilion)

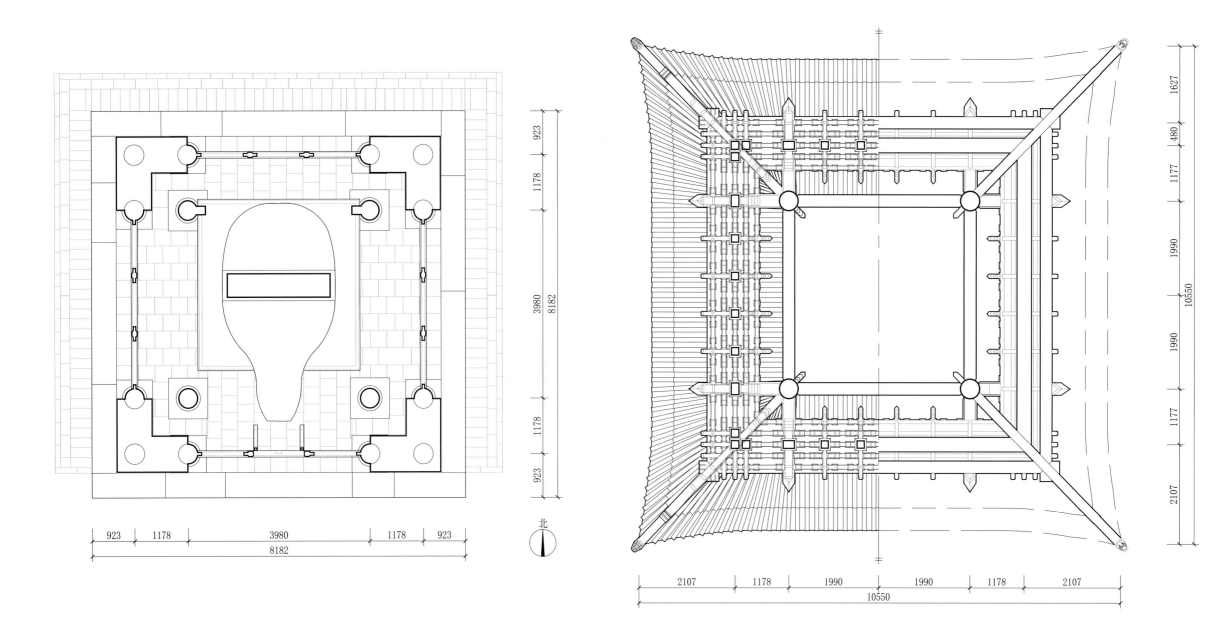

北

清乾隆十三年碑碑亭（五号碑亭）平面图
Plan of stele pavilion of the thirteenth reign year of emperor Qianlong,Qing Dynasty (No.5 stele pavilion)

清乾隆十三年碑碑亭（五号碑亭）下檐梁架仰俯视图
Plan of lower eaves framework of stele pavilion of the thirteenth reign year of emperor Qianlong,Qing Dynasty (No.5 stele pavilion), as seen from below

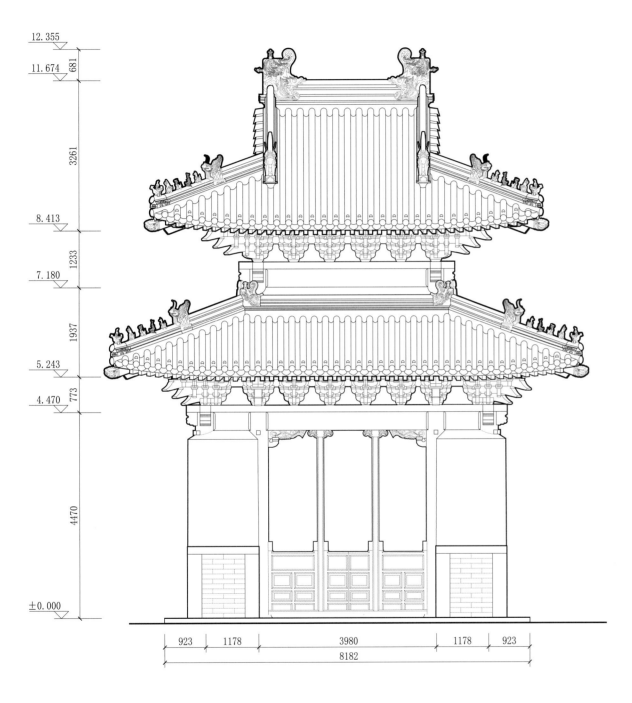

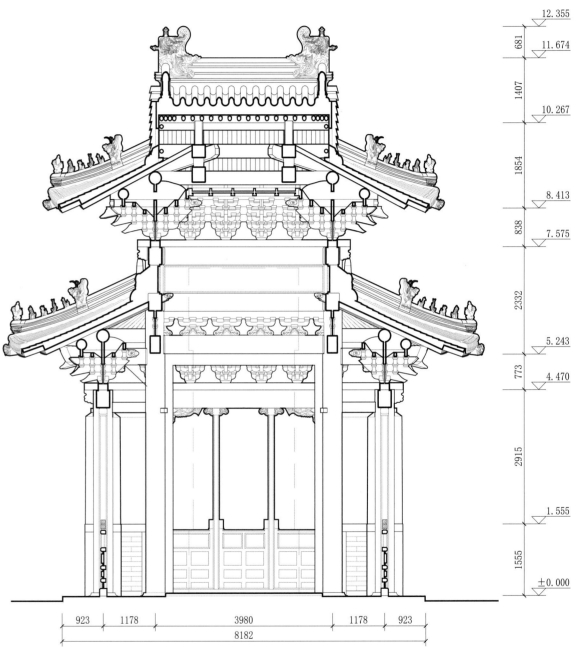

清乾隆十三年碑碑亭（五号碑亭）正立面图
Front elevation of stele pavilion of the thirteenth reign year of emperor Qianlong,Qing Dynasty (No.5 stele pavilion)

清乾隆十三年碑碑亭（五号碑亭）纵剖面图
Longitudinal section of stele pavilion of the thirteenth reign year of emperor Qianlong,Qing Dynasty (No.5 stele pavilion)

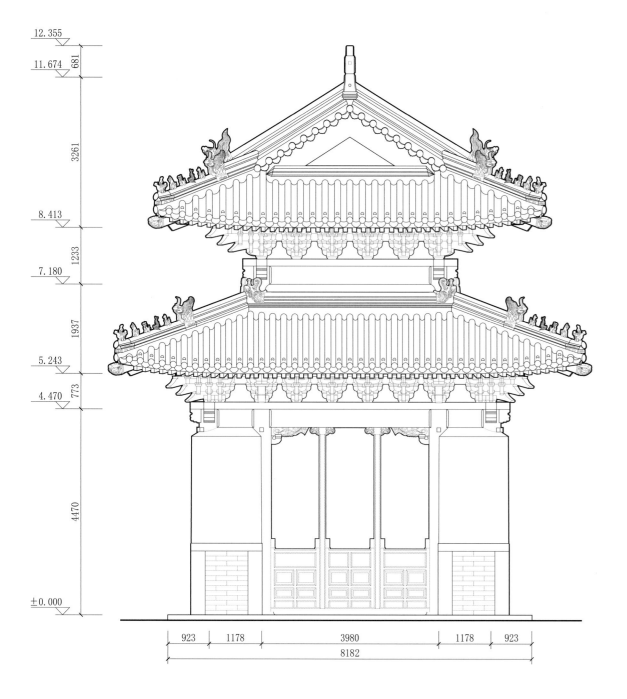

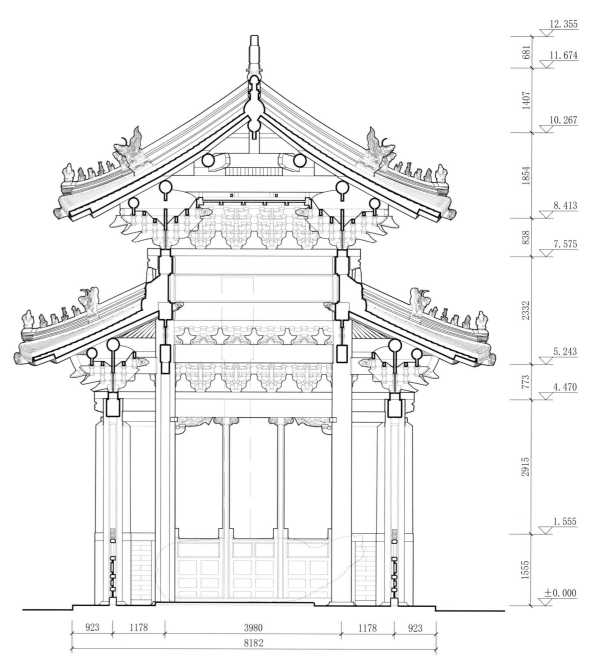

清乾隆十三年碑碑亭（五号碑亭）侧立面图
Side elevation of stele pavilion of the thirteenth reign year of emperor Qianlong,Qing Dynasty (No.5 stele pavilion)

清乾隆十三年碑碑亭（五号碑亭）横剖面图
Cross-section of stele pavilion of the thirteenth reign year of emperor Qianlong,Qing Dynasty (No.5 stele pavilion)

清康熙十五年碑碑亭（六号碑亭）
Stele pavilion of the fifteenth reign year of emperor Kangxi,Qing Dynasty (No.6 stele pavilion)

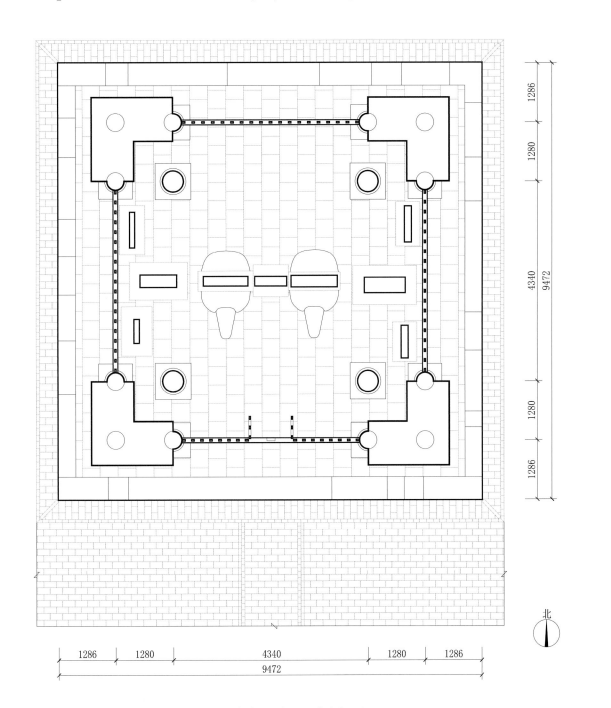

清康熙十五年碑碑亭（六号碑亭）平面图
Plan of stele pavilion of the fifteenth reign year of emperor Kangxi,Qing Dynasty (No.6 stele pavilion)

北

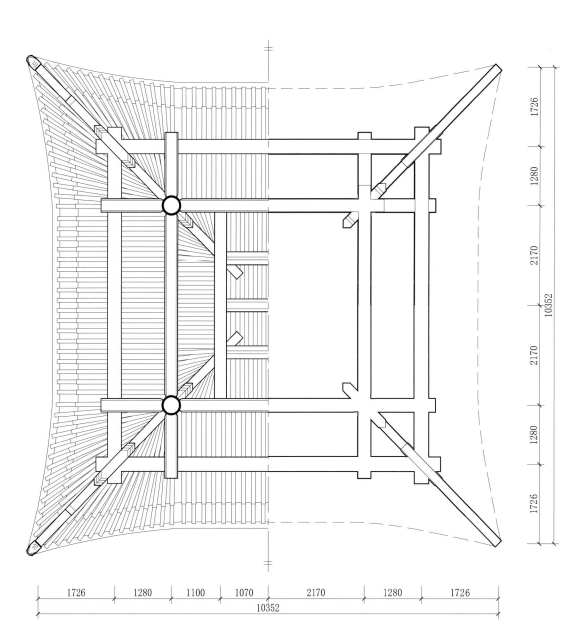

清康熙十五年碑碑亭（六号碑亭）下檐梁架仰俯视图
Plan of lower eaves framework of stele pavilion of the fifteenth reign year of emperor Kangxi,Qing Dynasty (No.6 stele pavilion), as seen from below

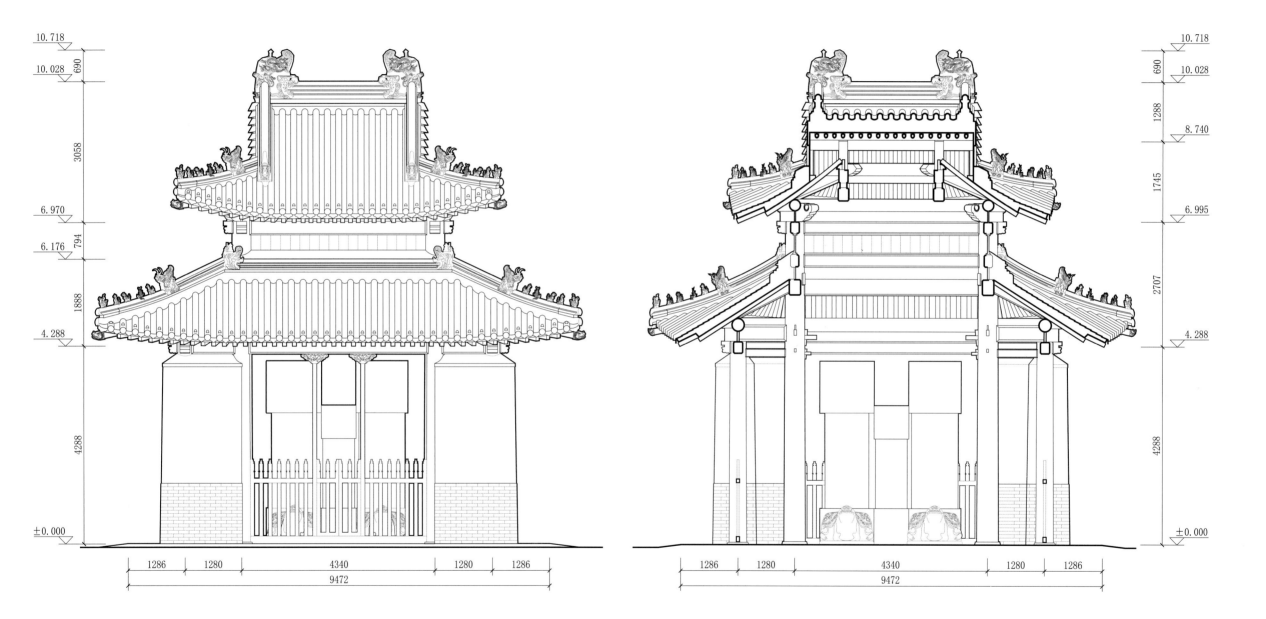

清康熙十五年碑碑亭（六号碑亭）正立面图

Front elevation of stele pavilion of the fifteenth reign year of emperor Kangxi,Qing Dynasty (No.6 stele pavilion)

清康熙十五年碑碑亭（六号碑亭）纵剖面图

Longitudinal section of stele pavilion of the fifteenth reign year of emperor Kangxi,Qing Dynasty (No.6 stele pavilion)

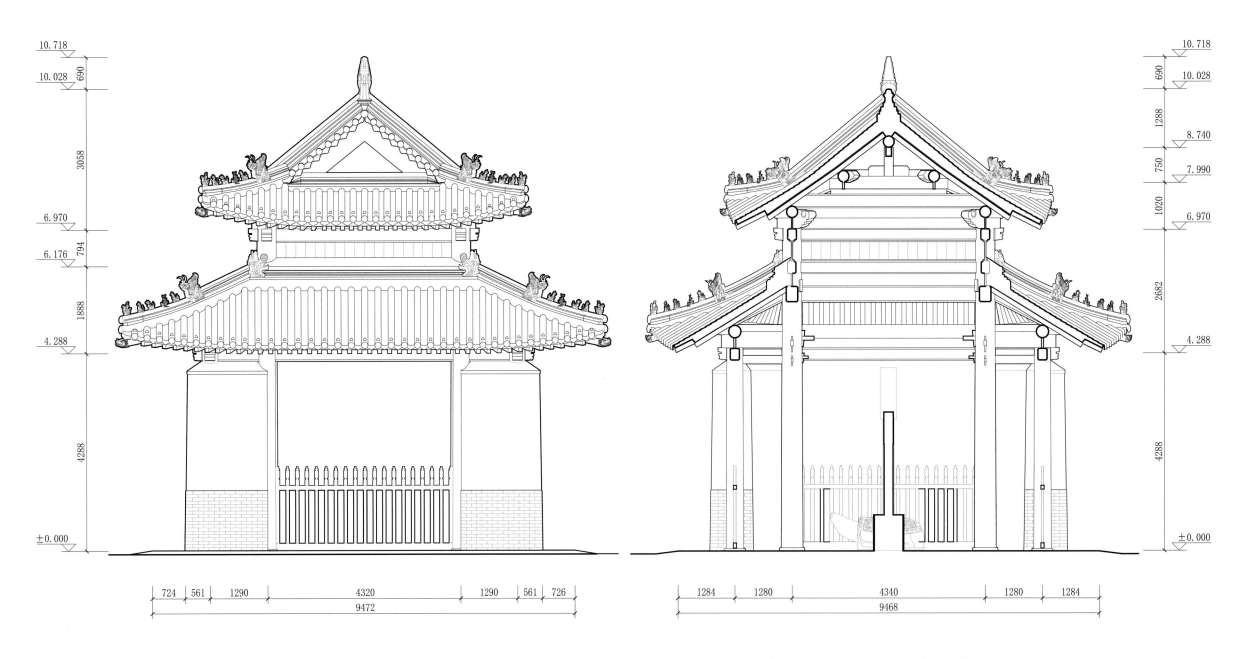

清康熙十五年碑碑亭（六号碑亭）侧立面图
Side elevation of stele pavilion of the fifteenth reign year of emperor Kangxi,Qing Dynasty (No.6 stele pavilion)

清康熙十五年碑碑亭（六号碑亭）横剖面图
Cross-section of stele pavilion of the fifteenth reign year of emperor Kangxi,Qing Dynasty (No.6 stele pavilion)

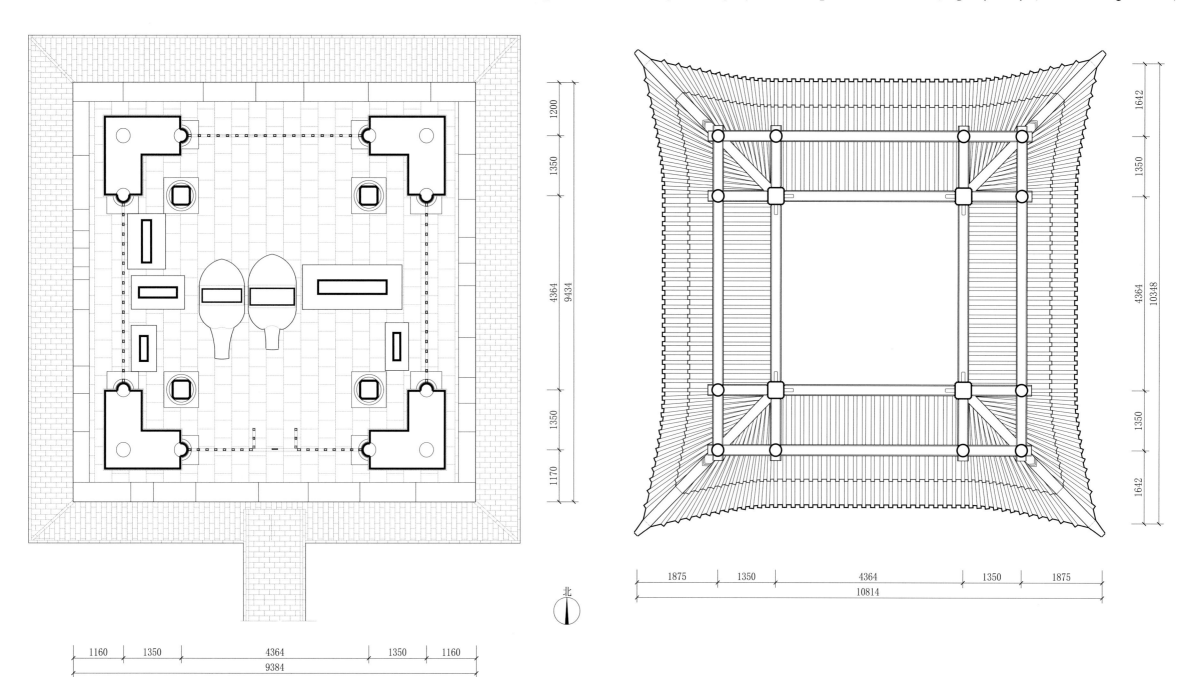

北

清顺治八年碑碑亭（七号碑亭）平面图
Plan of stele pavilion of the eighth reign year of emperor Shunzhi,Qing Dynasty (No.7 stele pavilion)

清顺治八年碑碑亭（七号碑亭）下檐梁架仰俯视图
Plan of lower eaves framework of stele pavilion of the eighth reign year of emperor Shunzhi,Qing Dynasty (No.7 stele pavilion), as seen from below

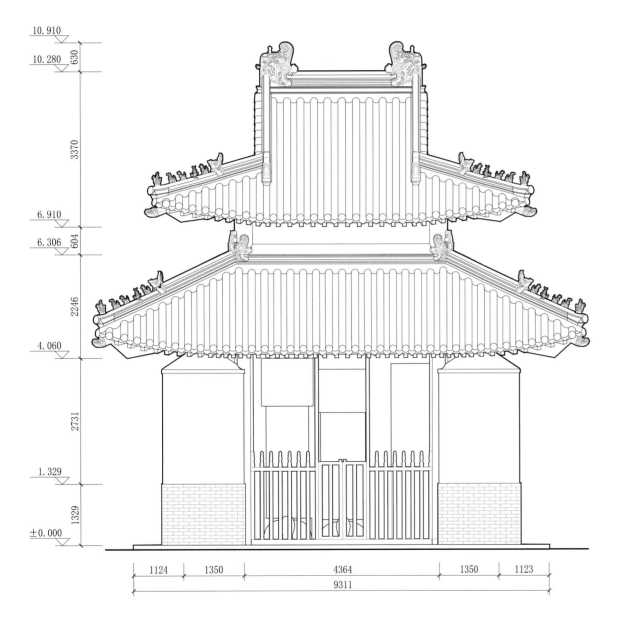

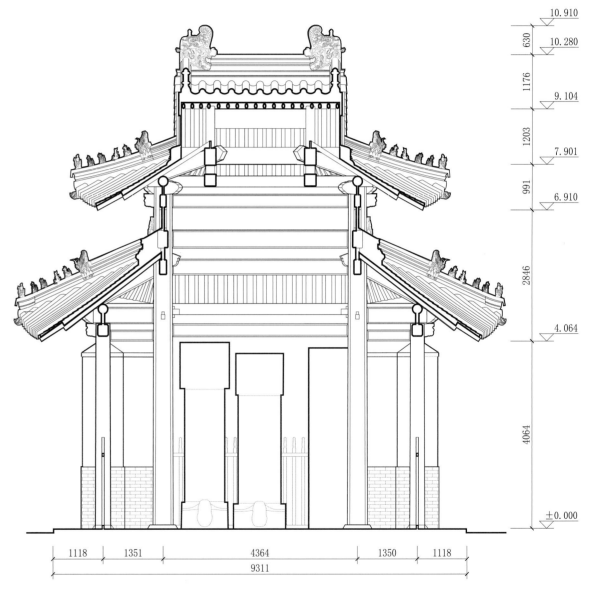

清顺治八年碑碑亭（七号碑亭）正立面图
Front elevation of stele pavilion of the eighth reign year of emperor Shunzhi,Qing Dynasty (No.7 stele pavilion)

清顺治八年碑碑亭（七号碑亭）纵剖面图
Longitudinal section of stele pavilion of the eighth reign year of emperor Shunzhi,Qing Dynasty (No.7 stele pavilion)

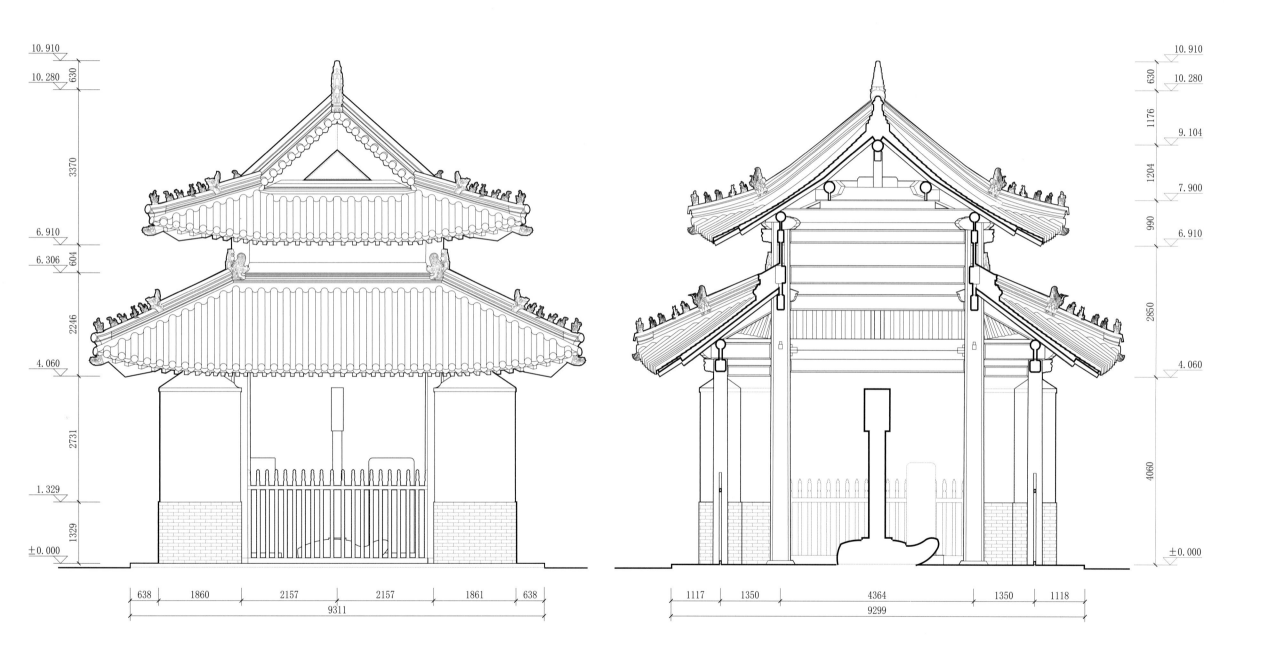

清顺治八年碑碑亭（七号碑亭）侧立面图
Side elevation of stele pavilion of the eighth reign year of emperor Shunzhi,Qing Dynasty (No.7 stele pavilion)

清顺治八年碑碑亭（七号碑亭）横剖面图
Cross-section of stele pavilion of the eighth reign year of emperor Shunzhi,Qing Dynasty (No.7 stele pavilion)

唐开元七年碑碑亭（八号碑亭）
Stele pavilion of the seventh reign year of Kaiyuan period, Tang Dynasty (No.8 stele pavilion)

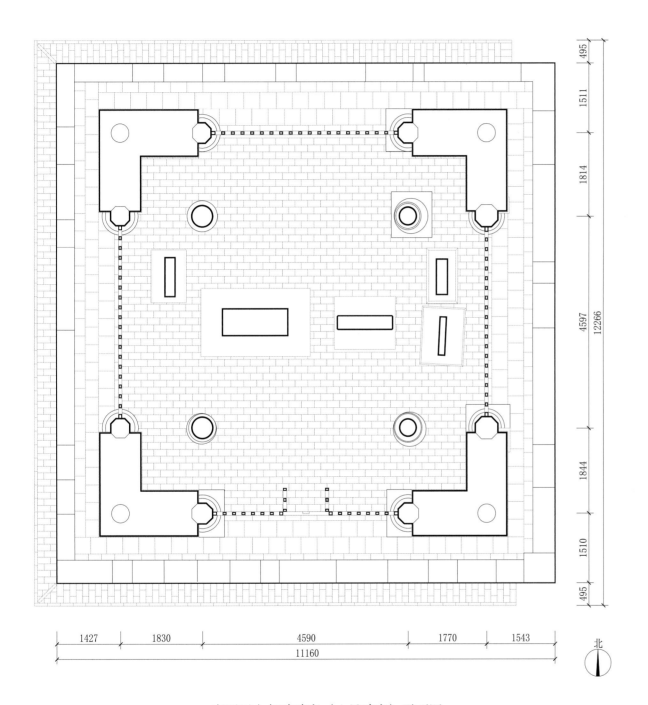

北

唐开元七年碑碑亭（八号碑亭）平面图
Plan of stele pavilion of the seventh reign year of Kaiyuan period, Tang Dynasty (No.8 stele pavilion)

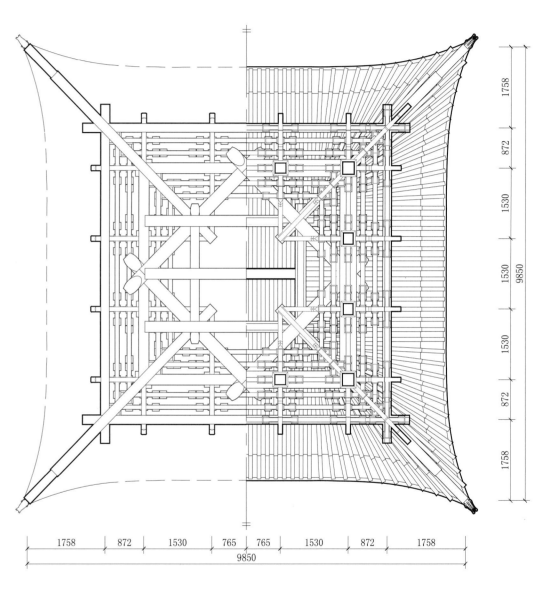

唐开元七年碑碑亭（八号碑亭）上檐梁架仰俯视图
Plan of upper eaves framework of stele pavilion of the seventh reign year of Kaiyuan period, Tang Dynasty (No.8 stele pavilion), as seen from below

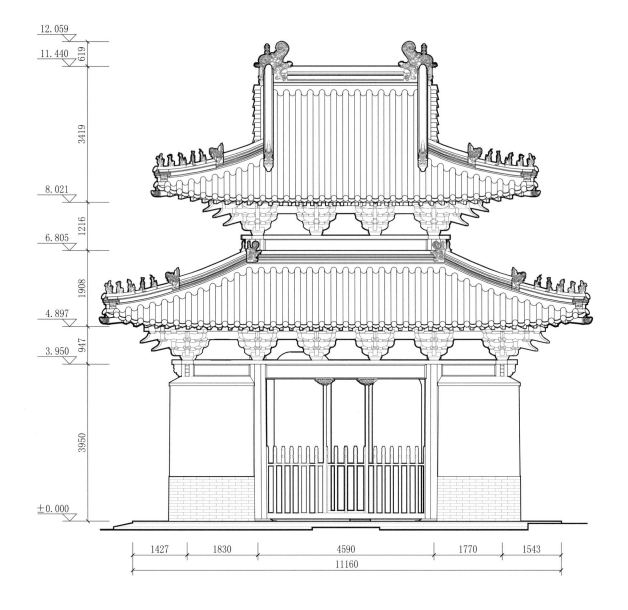

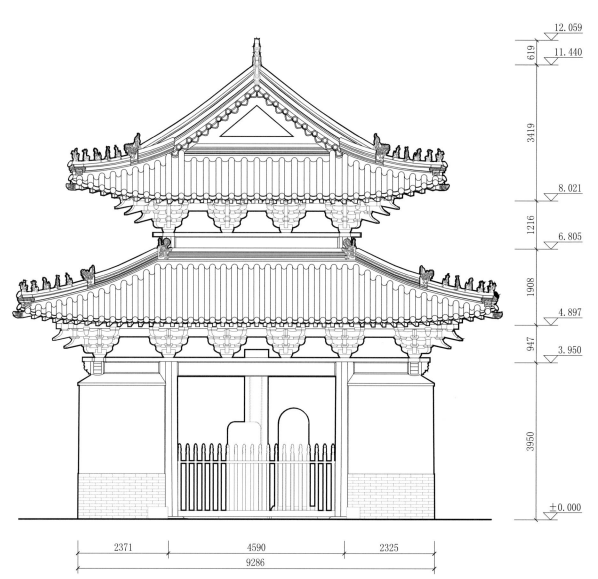

唐开元七年碑碑亭（八号碑亭）正立面图
Front elevation of stele pavilion of the seventh reign year of Kaiyuan period, Tang Dynasty (No.8 stele pavilion)

唐开元七年碑碑亭（八号碑亭）侧立面图
Side elevation of stele pavilion of the seventh reign year of Kaiyuan period, Tang Dynasty (No.8 stele pavilion)

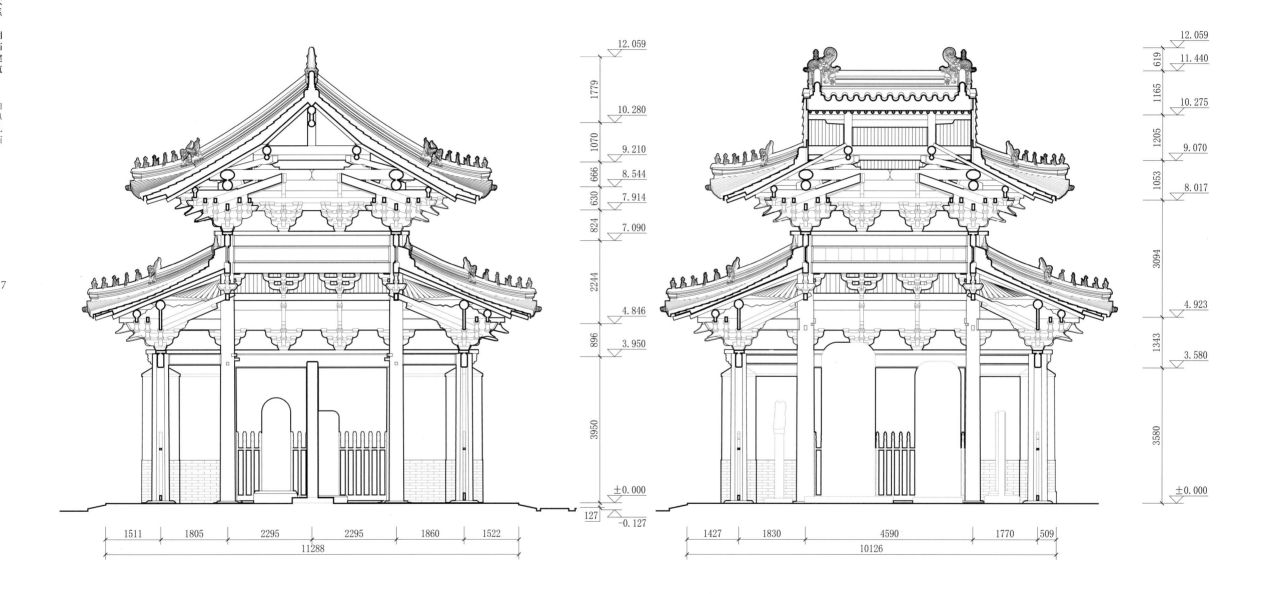

唐开元七年碑碑亭（八号碑亭）横剖面图
Cross-section of stele pavilion of the seventh reign year of Kaiyuan period,Tang Dynasty (No.8 stele pavilion)

唐开元七年碑碑亭（八号碑亭）纵剖面图
Longitudinal section of stele pavilion of the seventh reign year of Kaiyuan period,Tang Dynasty (No.8 stele pavilion)

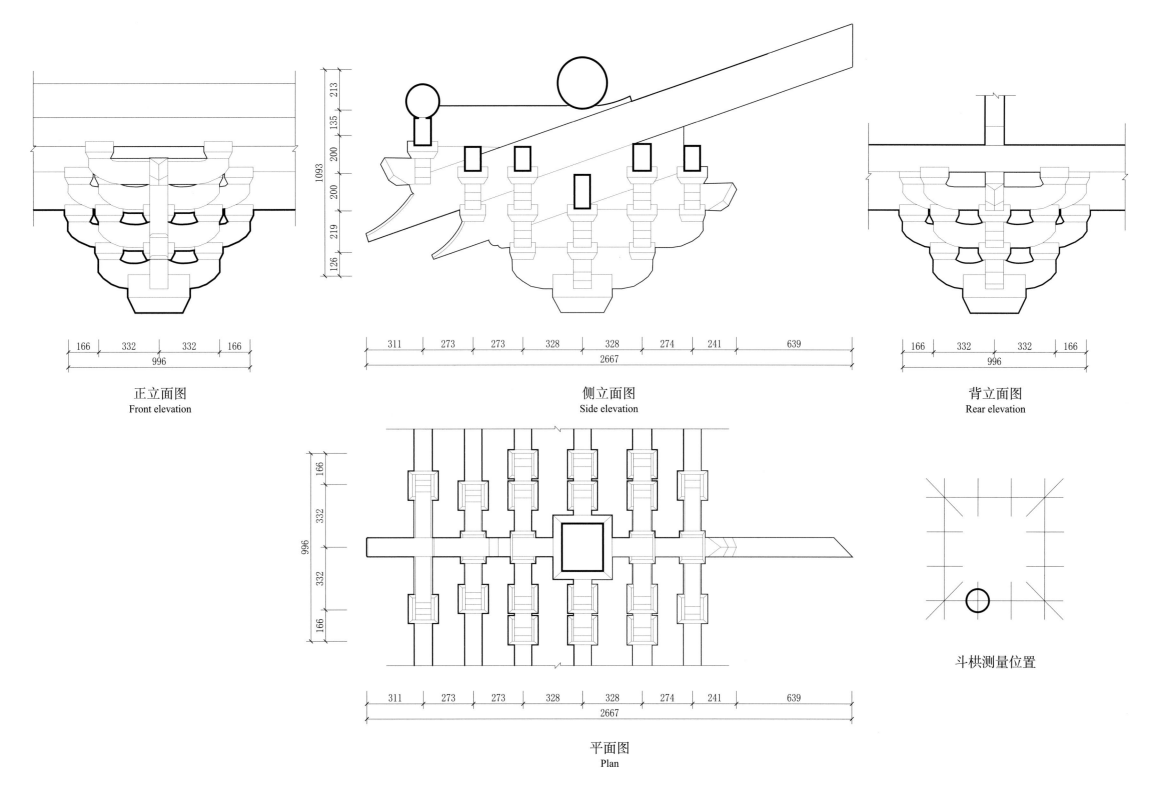

正立面图
Front elevation

侧立面图
Side elevation

背立面图
Rear elevation

平面图
Plan

斗栱测量位置

唐开元七年碑碑亭（八号碑亭）上檐平身科斗栱大样
Intercolumnar bracket sets of upper eaves of stele pavilion of the seventh reign year of Kaiyuan period,Tang Dynasty（No.8 stele pavilion）

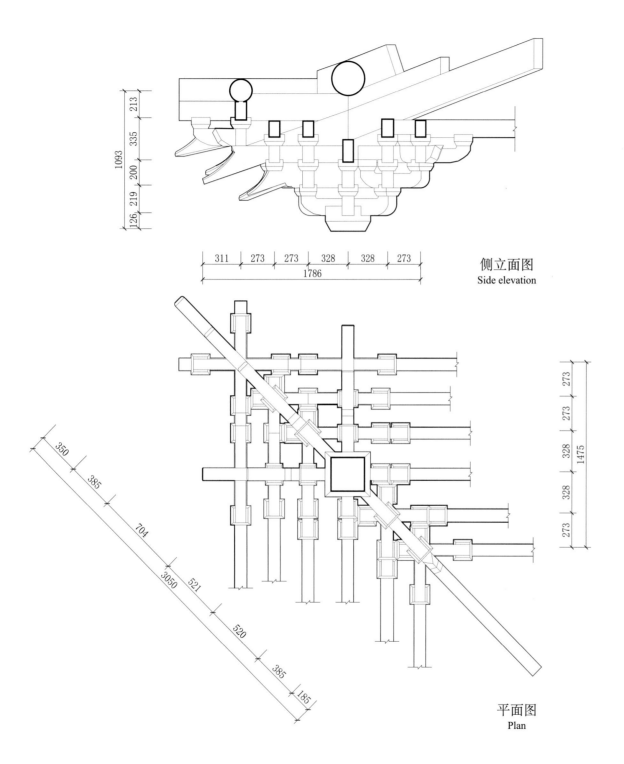

1093　213　335　200　219　126

311　273　273　328　328　273
1786

侧立面图
Side elevation

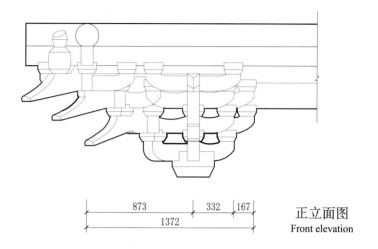

873　332　167
1372

正立面图
Front elevation

273　273　328　328　273
1475

350　385　704　521　520　385　185
3050

平面图
Plan

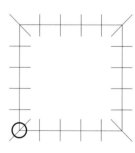

斗栱测量位置

唐开元七年碑碑亭（八号碑亭）上檐角科斗栱大样
Corner bracket sets of stele pavilion of the seventh reign year of Kaiyuan period, Tang Dynasty （No.8 stele pavilion）

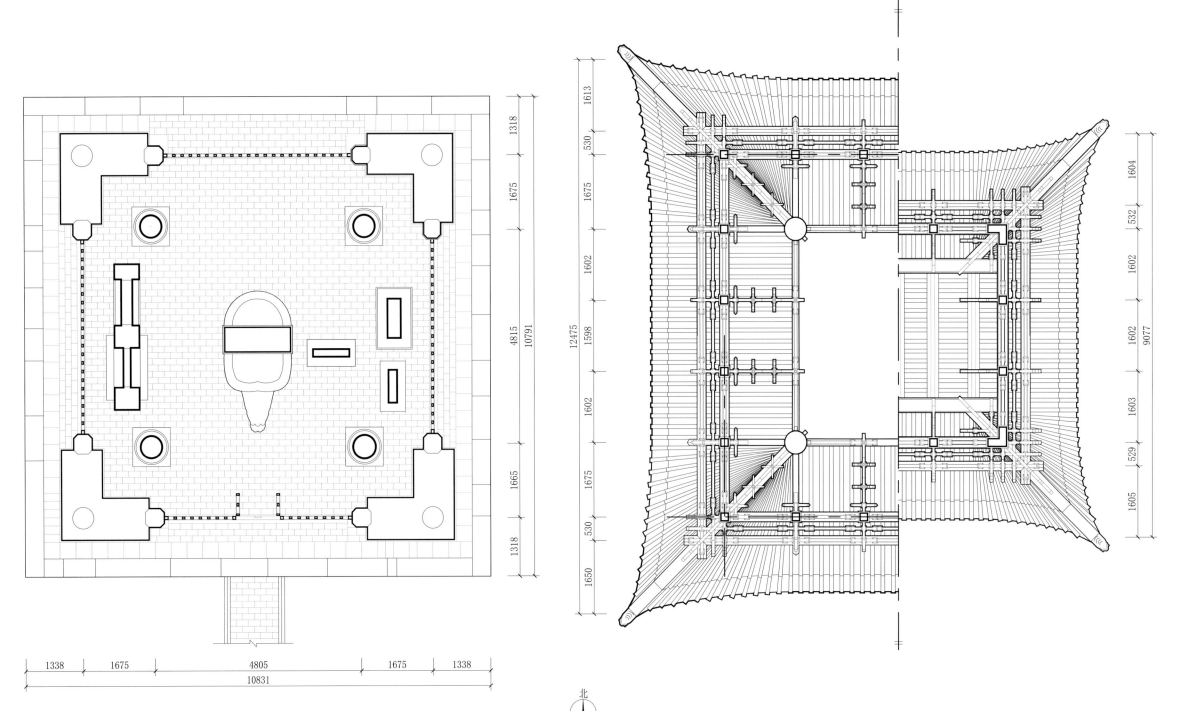

元至顺二年碑碑亭（九号碑亭）平面图
Plan of stele pavilion of the second reign year of Zhishun period, Yuan Dynasty (No.9 stele pavilion)

北

元至顺二年碑碑亭（九号碑亭）梁架仰视图
Plan of framework of stele pavilion of the second reign year of Zhishun period, Yuan Dynasty (No.9 stele pavilion), as seen from below

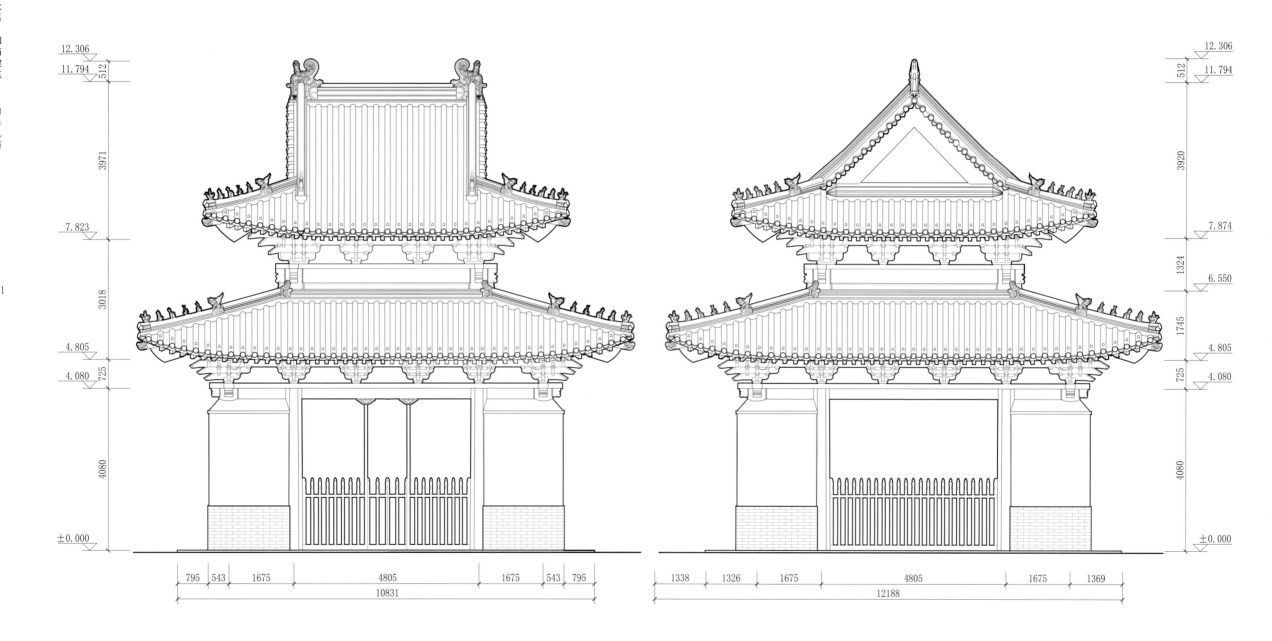

元至顺二年碑碑亭（九号碑亭）正立面图
Front elevation of stele pavilion of the second reign year of Zhishun period, Yuan Dynasty (No.9 stele pavilion)

元至顺二年碑碑亭（九号碑亭）侧立面图
Side elevation of stele pavilion of the second reign year of Zhishun period, Yuan Dynasty (No.9 stele pavilion)

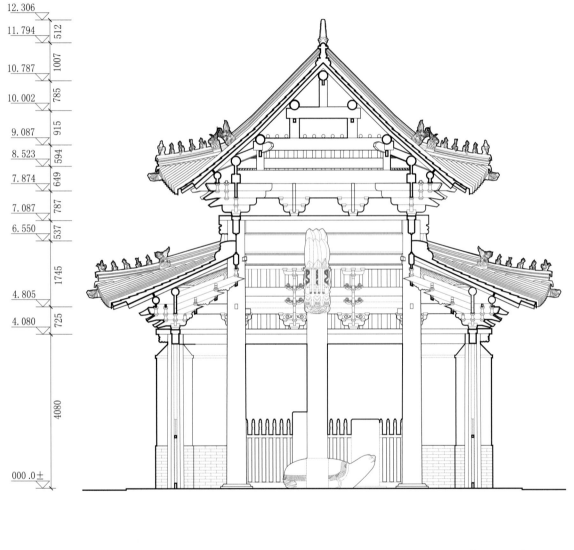

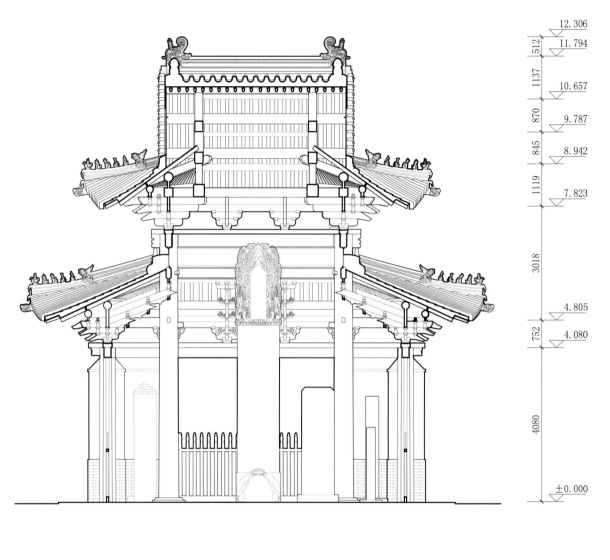

元至顺二年碑碑亭（九号碑亭）横剖面图
Cross-section of stele pavilion of the second reign year of Zhishun period,Yuan Dynasty (No.9 stele pavilion)

元至顺二年碑碑亭（九号碑亭）纵剖面图
Longitudinal section of stele pavilion of the second reign year of Zhishun period,Yuan Dynasty (No.9 stele pavilion)

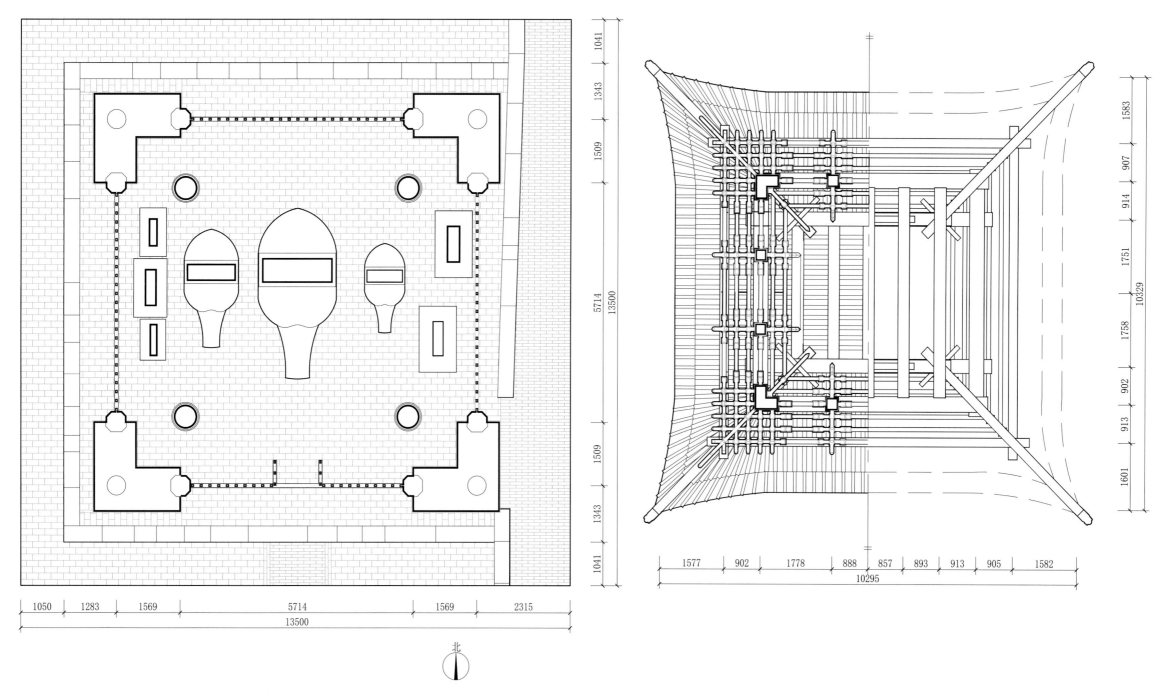

北

元至元五年碑碑亭（十号碑亭）平面图
Plan of stele pavilion of the fifth reign year of Zhiyuan period, Yuan Dynasty (No.10 stele pavilion)

元至元五年碑碑亭（十号碑亭）上檐梁架仰视及俯视图
Plan of upper eaves framework of stele pavilion of the fifth reign year of Zhiyuan period, Yuan Dynasty (No.10 stele pavilion), as seen from below

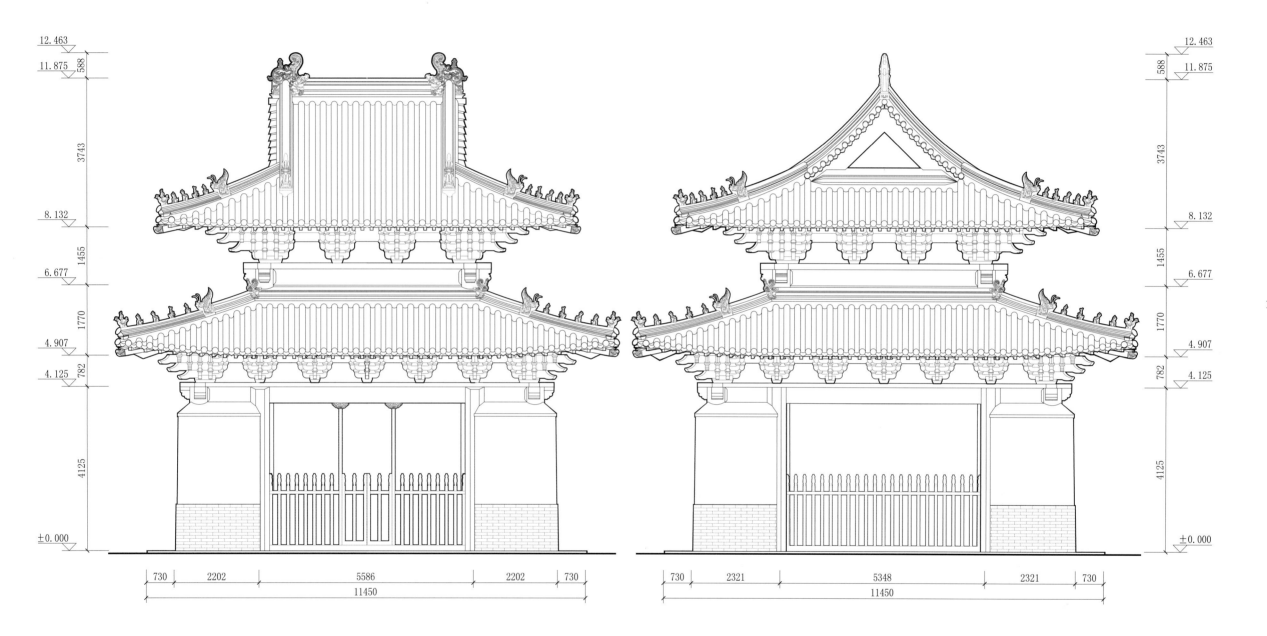

元至元五年碑碑亭（十号碑亭）正立面图
Front elevation of stele pavilion of the fifth reign year of Zhiyuan period, Yuan Dynasty (No.10 stele pavilion)

元至元五年碑碑亭（十号碑亭）侧立面图
Side elevation of stele pavilion of the fifth reign year of Zhiyuan period, Yuan Dynasty (No.10 stele pavilion)

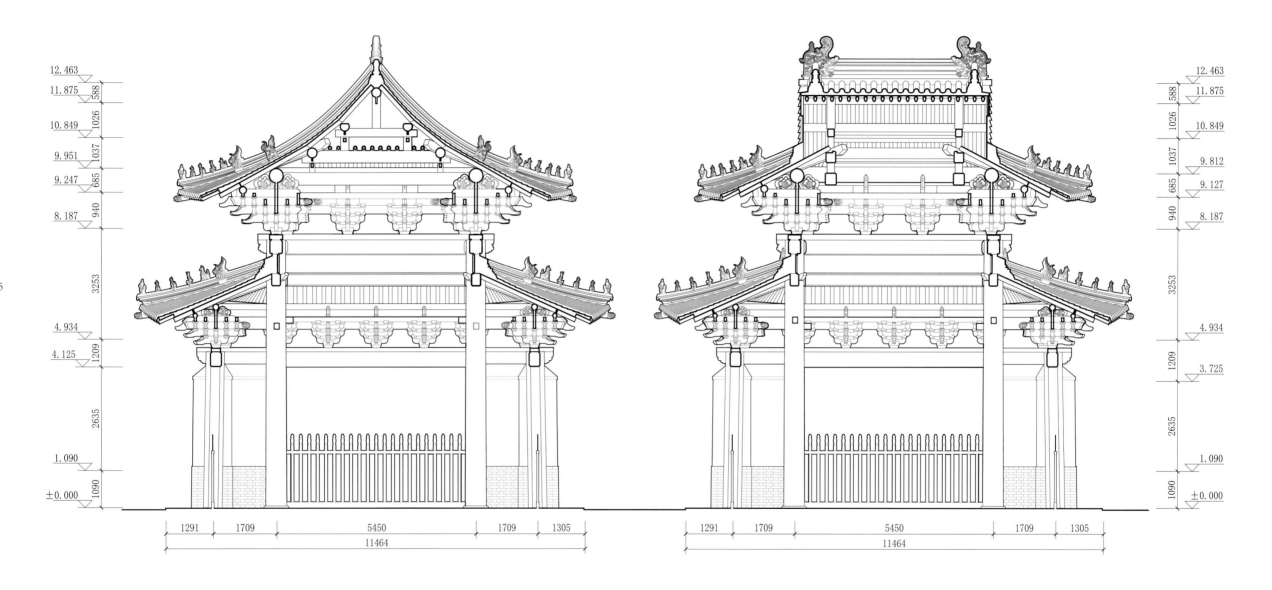

元至元五年碑碑亭（十号碑亭）横剖面图
Cross-section of stele pavilion of the fifth reign year of Zhiyuan period,Yuan Dynasty (No.10 stele pavilion)

元至元五年碑碑亭（十号碑亭）纵剖面图
Longitudinal section of stele pavilion of the fifth reign year of Zhiyuan period,Yuan Dynasty (No.10 stele pavilion)

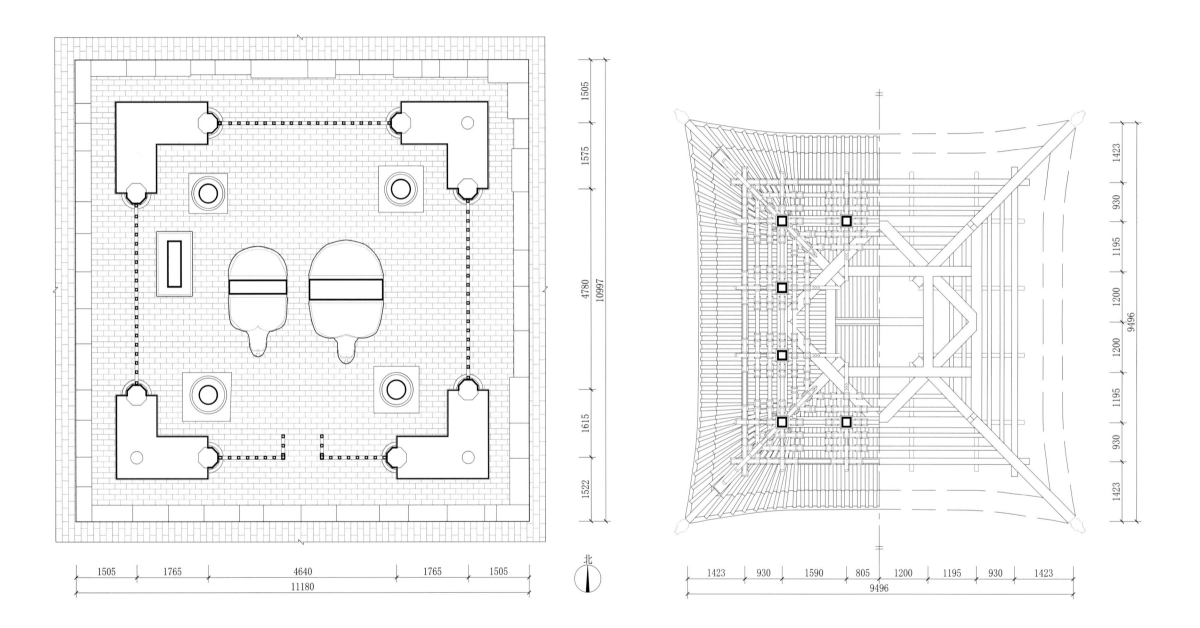

宋太平兴国八年碑碑亭（十一号碑亭）平面图

Plan of stele pavilion of the eighth reign year of "Taiping xingguo" period,Song Dynasty (No.11 stele pavilion)

宋太平兴国八年碑碑亭（十一号碑亭）上檐梁架仰俯视图

Plan of upper eaves framework of stele pavilion of the eighth reign year of "Taiping xingguo" period,Song Dynasty (No.11 stele pavilion), as seen from below

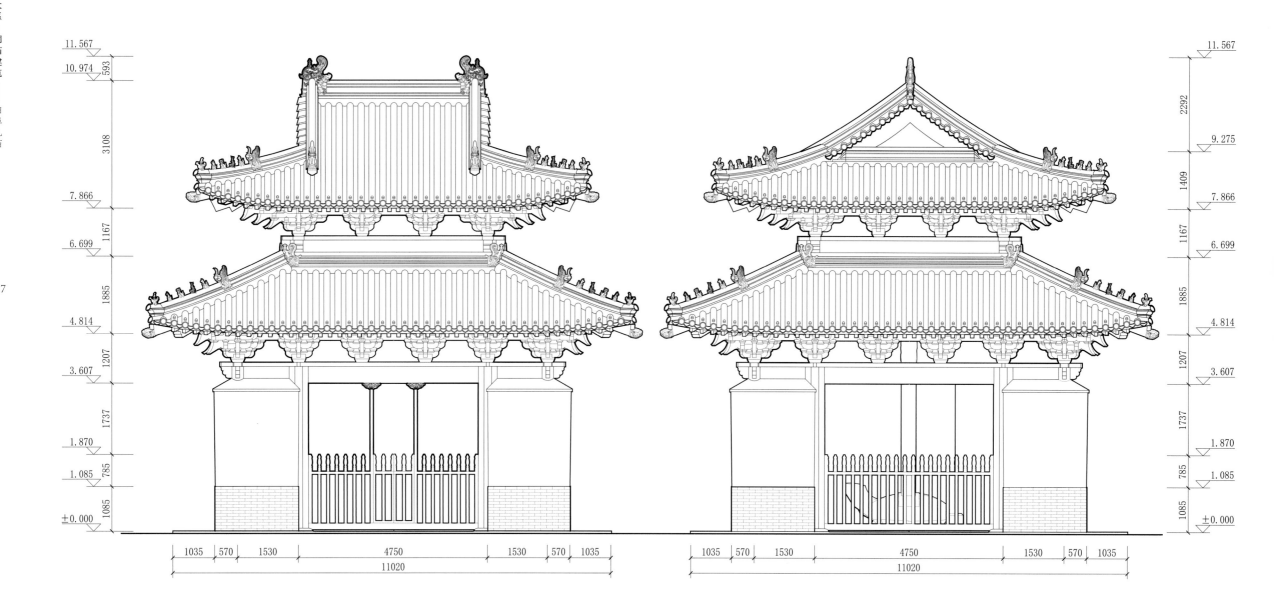

宋太平兴国八年碑碑亭（十一号碑亭）正立面图
Front elevation of stele pavilion of the eighth reign year of "Taiping xingguo" period,Song Dynasty (No.11 stele pavilion)

宋太平兴国八年碑碑亭（十一号碑亭）侧立面图
Side elevation of stele pavilion of the eighth reign year of "Taiping xingguo" period,Song Dynasty (No.11 stele pavilion)

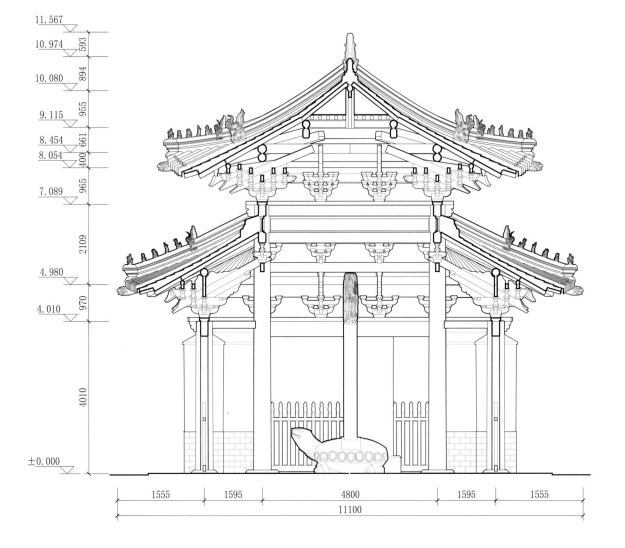

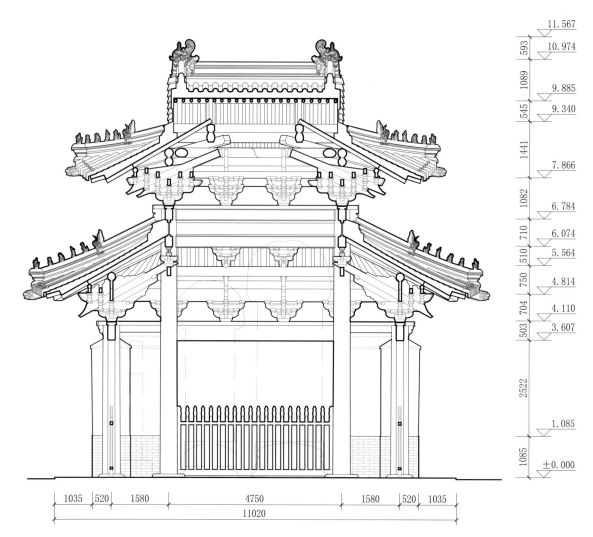

宋太平兴国八年碑碑亭（十一号碑亭）横剖面图
Cross-section of stele pavilion of the eighth reign year of "Taiping xingguo" period,Song Dynasty (No.11 stele pavilion)

宋太平兴国八年碑碑亭（十一号碑亭）纵剖面图
Longitudinal section of stele pavilion of the eighth reign year of "Taiping xingguo" period,Song Dynasty (No.11 stele pavilion)

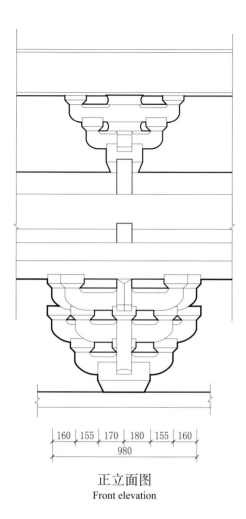

正立面图
Front elevation

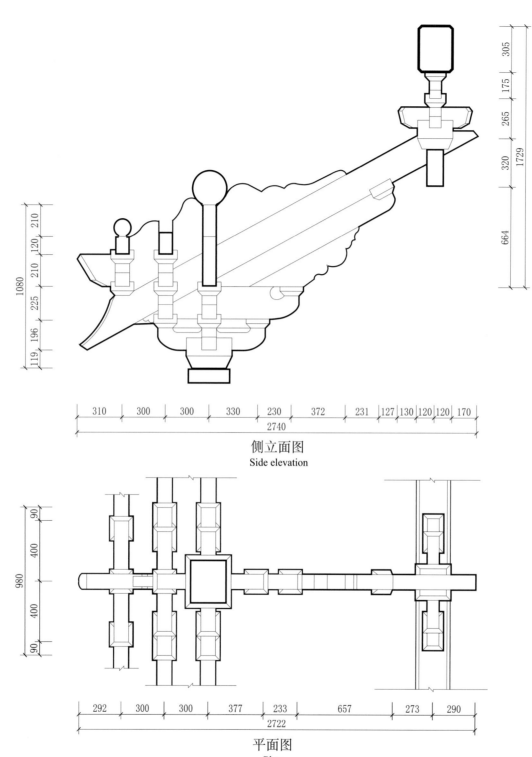

侧立面图
Side elevation

背立面图
Rear elevation

平面图
Plan

斗栱测量位置

宋太平兴国八年碑碑亭（十一号碑亭）下檐平身科斗栱
Intercolumnar bracket sets of lower eaves of stele pavilion of the eighth reign year of "Taiping xingguo" period, Song Dynasty (No.11 stele pavilion)

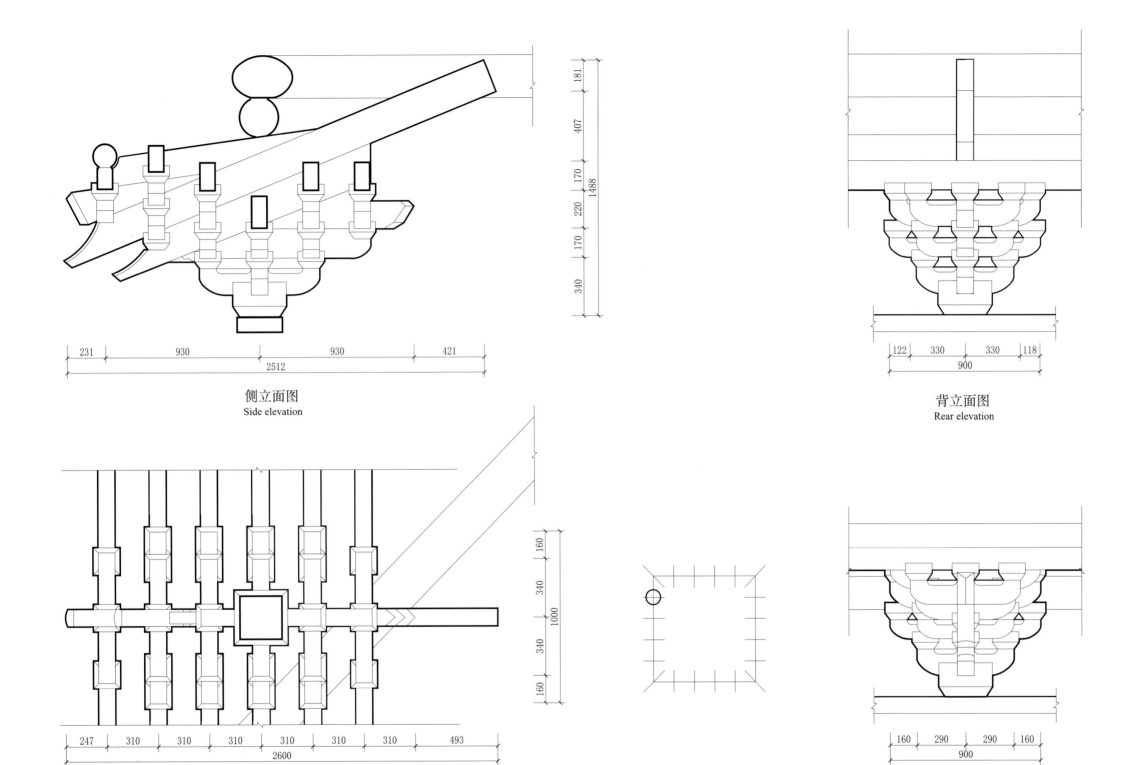

侧立面图
Side elevation

背立面图
Rear elevation

平面图
Plan

正立面图
Front elevation

宋太平兴国八年碑碑亭（十一号碑亭）上檐平身科斗栱大样
Intercolumnar bracket sets of upper eaves of stele pavilion of the eighth reign year of "Taiping xingguo" period, Song Dynasty (No.11 stele pavilion)

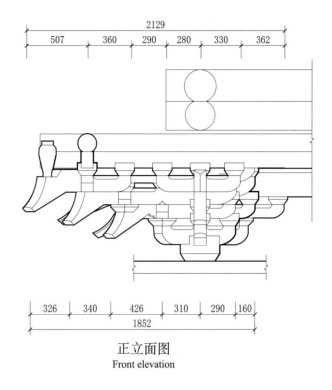

2910
202 360 310 310 310 310 310 238 560

250 260 360 310 340 1520

侧立面图
Side elevation

2129
507 360 290 280 330 362

326 340 426 310 290 160
1852

正立面图
Front elevation

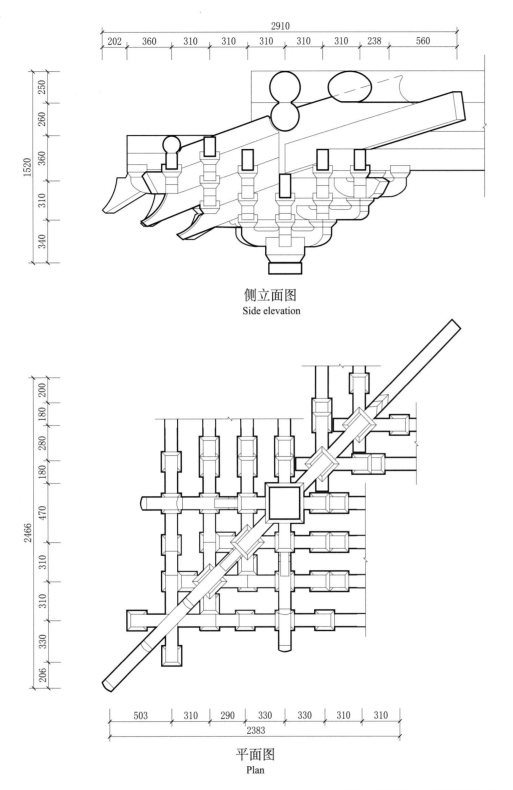

200 180 280 180 470 310 310 330 206
2466

503 310 290 330 330 310 310
2383

平面图
Plan

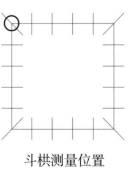

斗栱测量位置

宋太平兴国八年碑碑亭（十一号碑亭）上檐角科斗栱大样
Corner bracket sets of stele pavilion of the eighth reign year of "Taiping xingguo" period,Song Dynasty (No.11 stele pavilion)

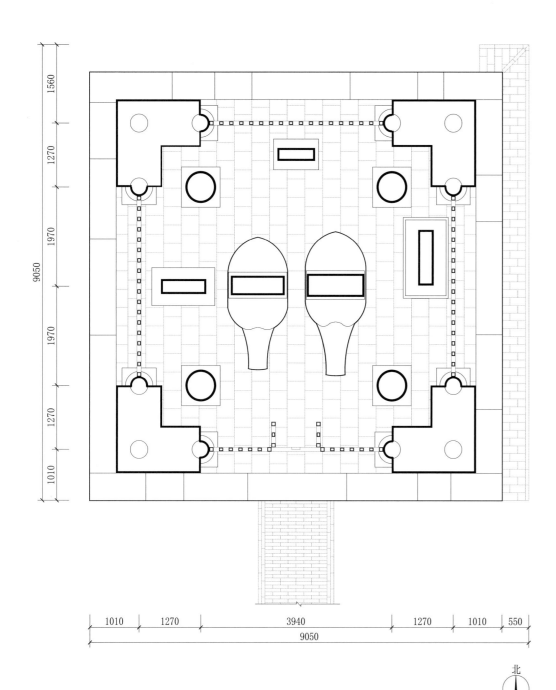

清雍正元年碑西碑亭（十二号碑亭）平面图

Plan of west stele pavilion of the first reign year of emperor Yongzheng,Qing Dynasty (No.12 stele pavilion)

北

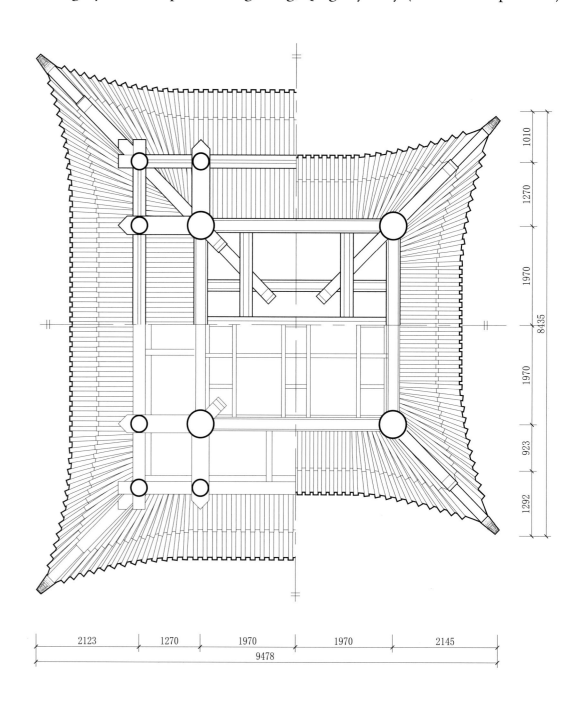

清雍正元年碑西碑亭（十二号碑亭）梁架仰视图

Plan of framework of west stele pavilion of the first reign year of emperor Yongzheng,Qing Dynasty (No.12 stele pavilion), as seen from below

133

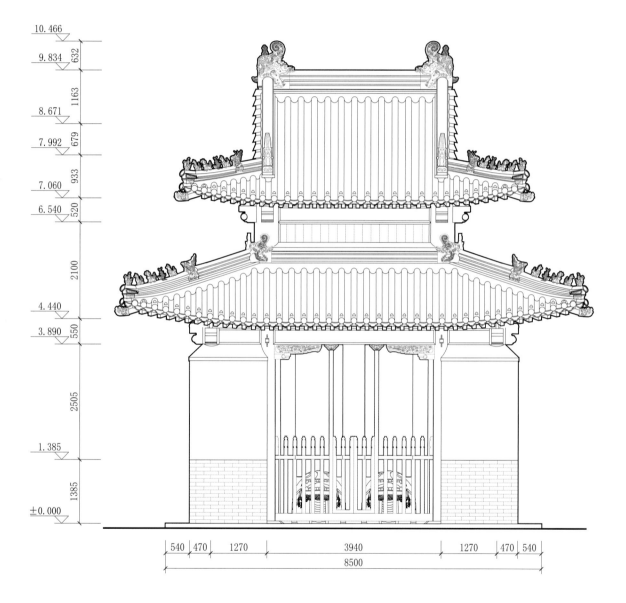

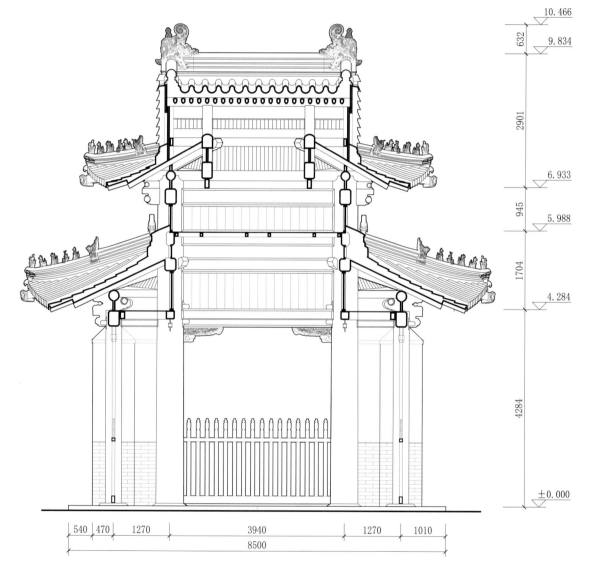

清雍正元年碑西碑亭（十二号碑亭）正立面图
Front elevation of west stele pavilion of the first reign year of emperor Yongzheng,Qing Dynasty (No.12 stele pavilion)

清雍正元年碑西碑亭（十二号碑亭）纵剖面图
Longitudinal section of west stele pavilion of the first reign year of emperor Yongzheng,Qing Dynasty (No.12 stele pavilion)

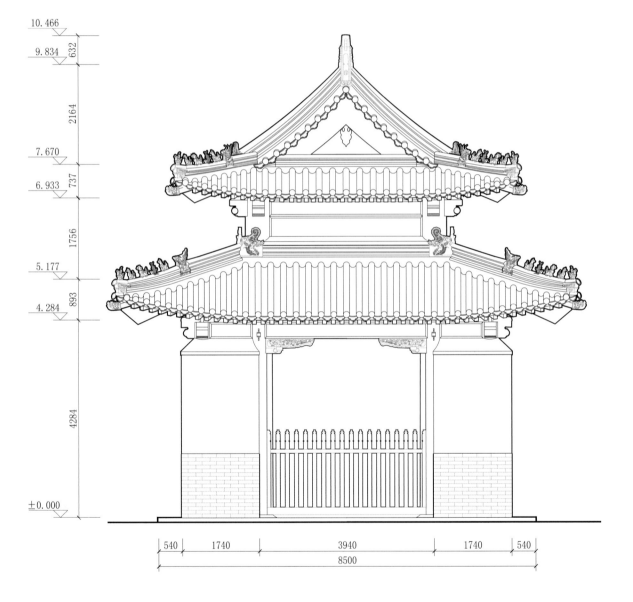

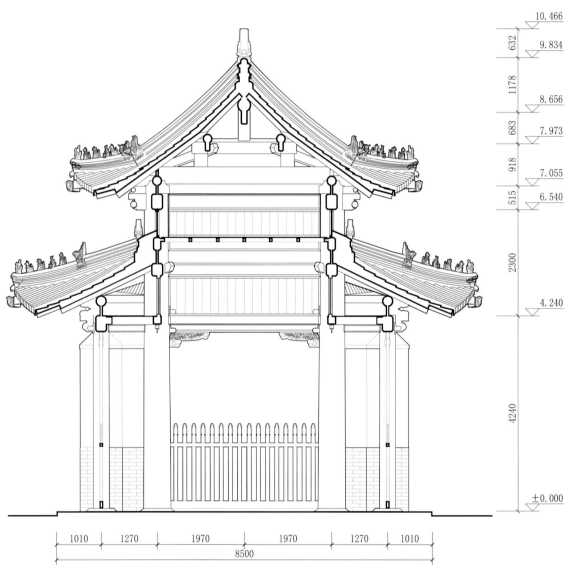

清雍正元年碑西碑亭（十二号碑亭）侧立面图
Side elevation of west stele pavilion of the first reign year of emperor Yongzheng,Qing Dynasty (No.12 stele pavilion)

清雍正元年碑西碑亭（十二号碑亭）横剖面图
Cross-section of west stele pavilion of the first reign year of emperor Yongzheng,Qing Dynasty (No.12 stele pavilion)

清雍正元年碑东碑亭（十三号碑亭）
East stele pavilion of the first reign year of emperor Yongzheng,Qing Dynasty (No.13 stele pavilion)

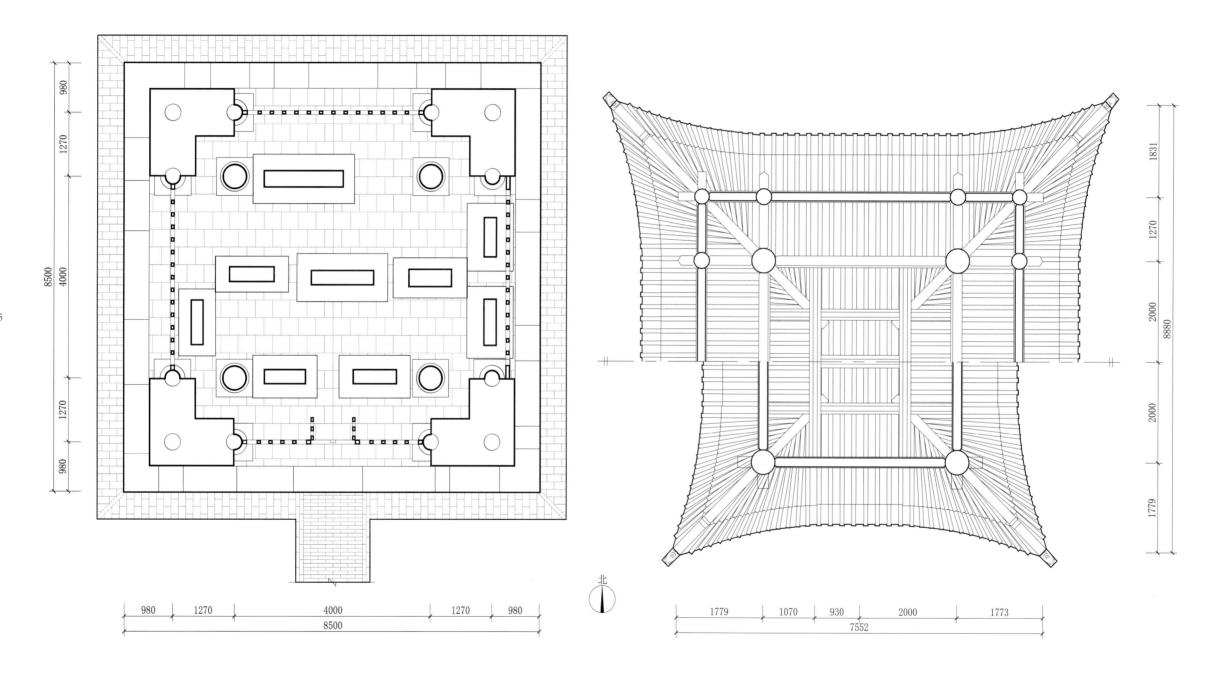

清雍正元年碑东碑亭（十三号碑亭）平面图
Plan of east stele pavilion of the first reign year of emperor Yongzheng,Qing Dynasty (No.13 stele pavilion)

清雍正元年碑东碑亭（十三号碑亭）梁架仰视图
Plan of framework of east stele pavilion of the first reign year of emperor Yongzheng,Qing Dynasty (No.13 stele pavilion), as seen from below

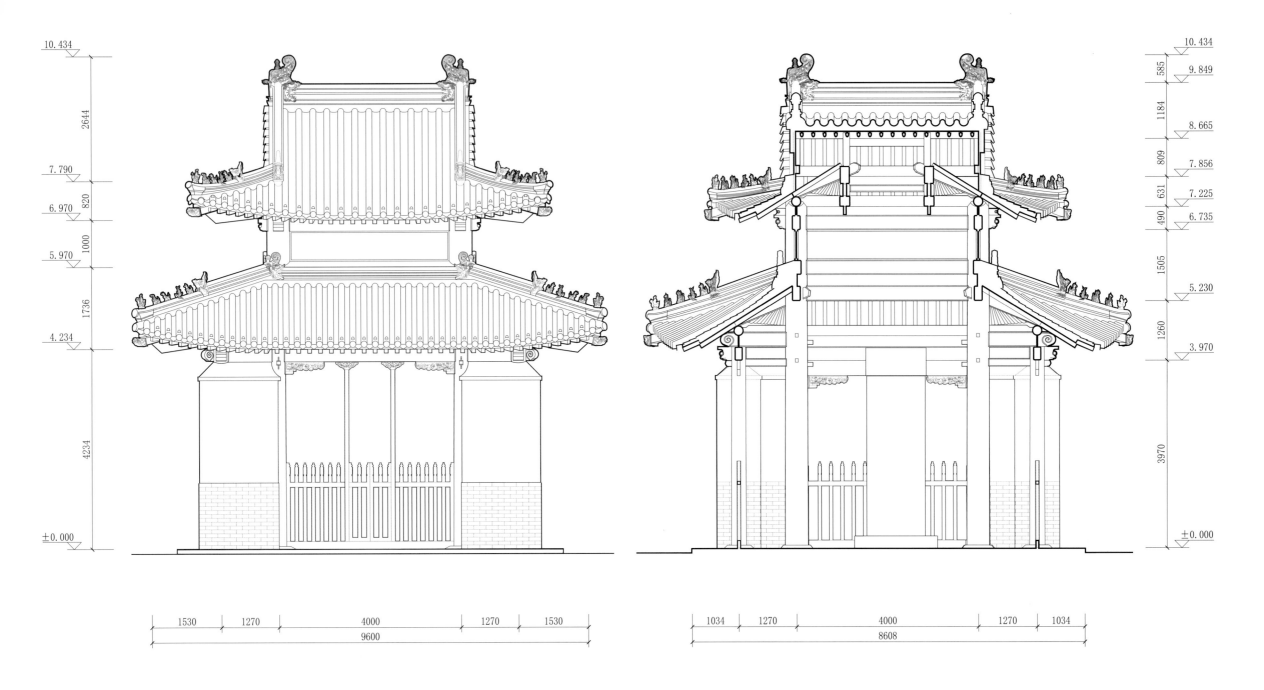

10.434

2644

7.790

6.970   820

5.970   1000

1736

4.234

4234

±0.000

1530 | 1270 | 4000 | 1270 | 1530
9600

清雍正元年碑东碑亭（十三号碑亭）正立面图
Front elevation of east stele pavilion of the first reign year of emperor Yongzheng,Qing Dynasty (No.13 stele pavilion)

585   10.434
9.849
1184
8.665
809   7.856
631   7.225
490   6.735
1505
5.230
1260
3.970

3970

±0.000

1034 | 1270 | 4000 | 1270 | 1034
8608

清雍正元年碑东碑亭（十三号碑亭）纵剖面图
Longitudinal section of east stele pavilion of the first reign year of emperor Yongzheng,Qing Dynasty (No.13 stele pavilion)

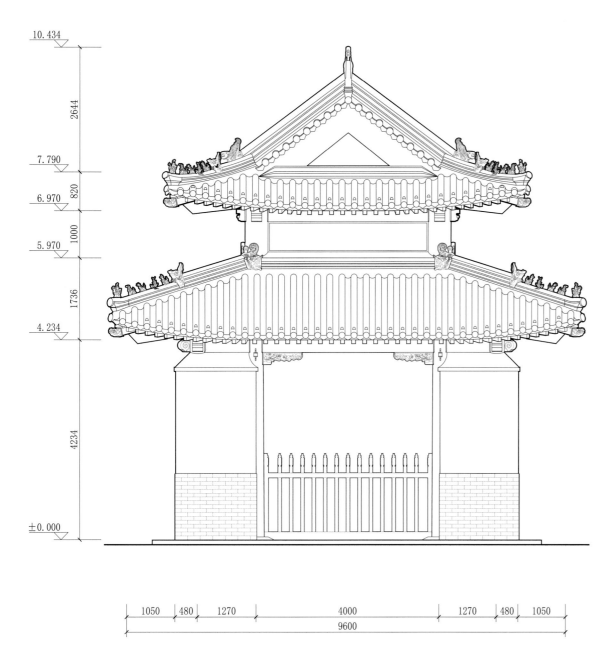

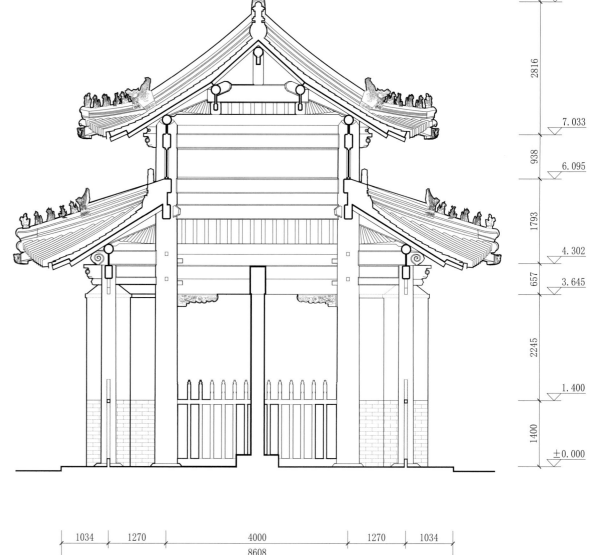

清雍正元年碑东碑亭（十三号碑亭）侧立面图
Side elevation of east stele pavilion of the first reign year of emperor Yongzheng,Qing Dynasty (No.13 stele pavilion)

清雍正元年碑东碑亭（十三号碑亭）横剖面图
Cross-section of east stele pavilion of the first reign year of emperor Yongzheng,Qing Dynasty (No.13 stele pavilion)

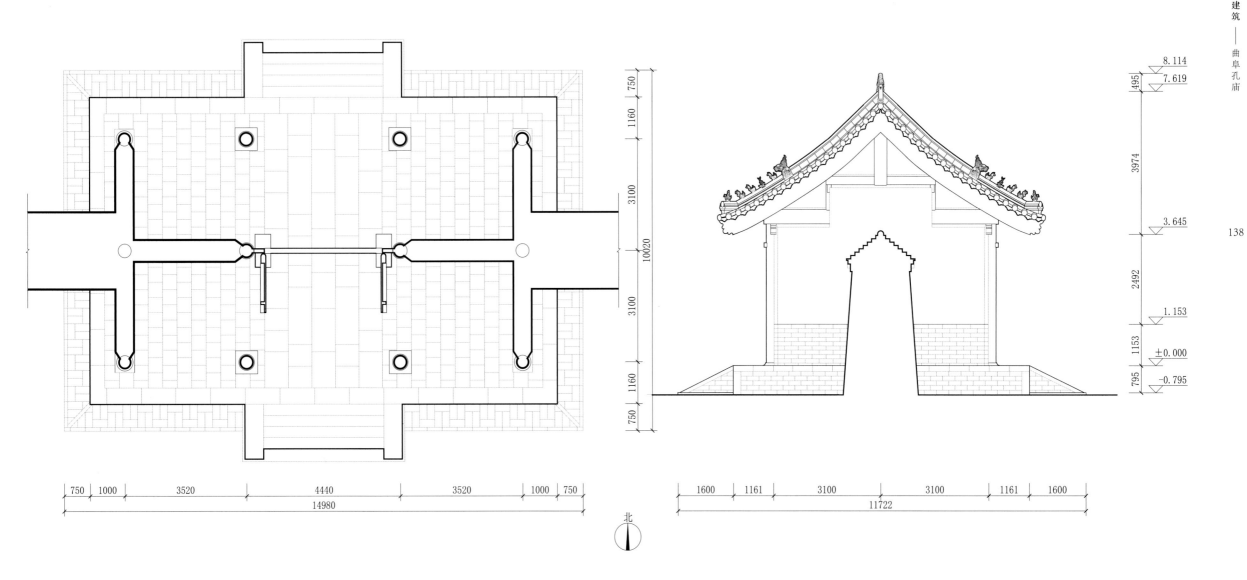

观德门平面图
Plan of Guande Gate

观德门侧立面图
Side elevation of Guande Gate

北

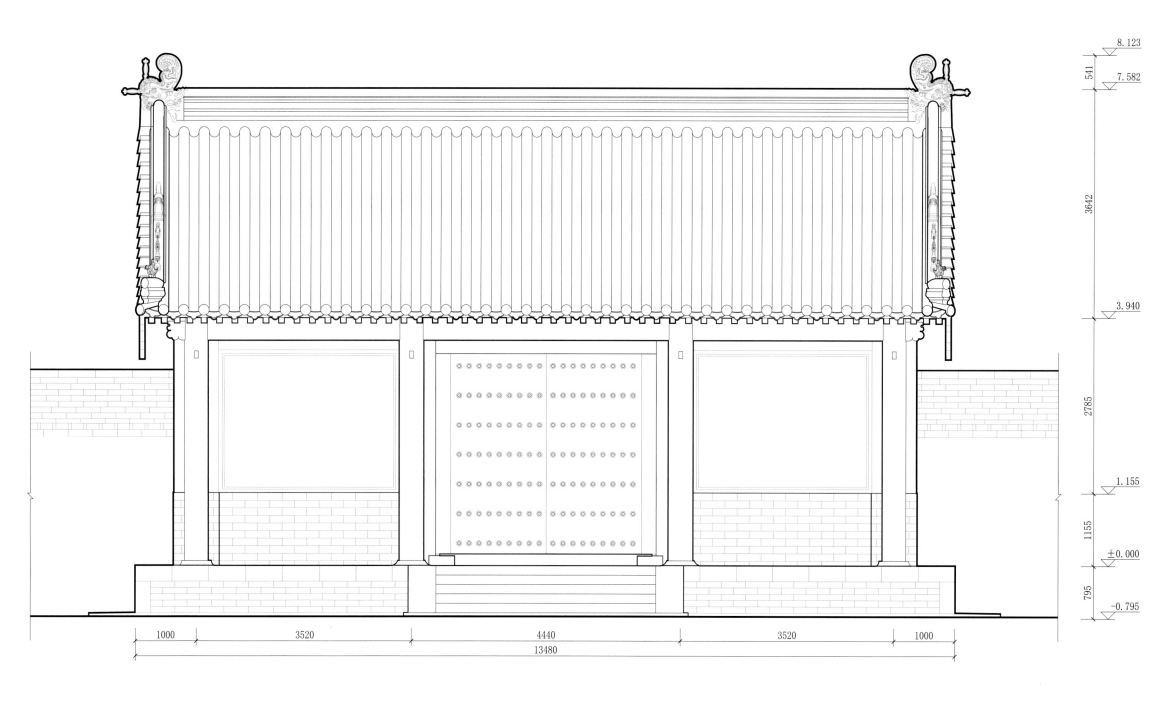

观德门正立面图
Front elevation of Guande Gate

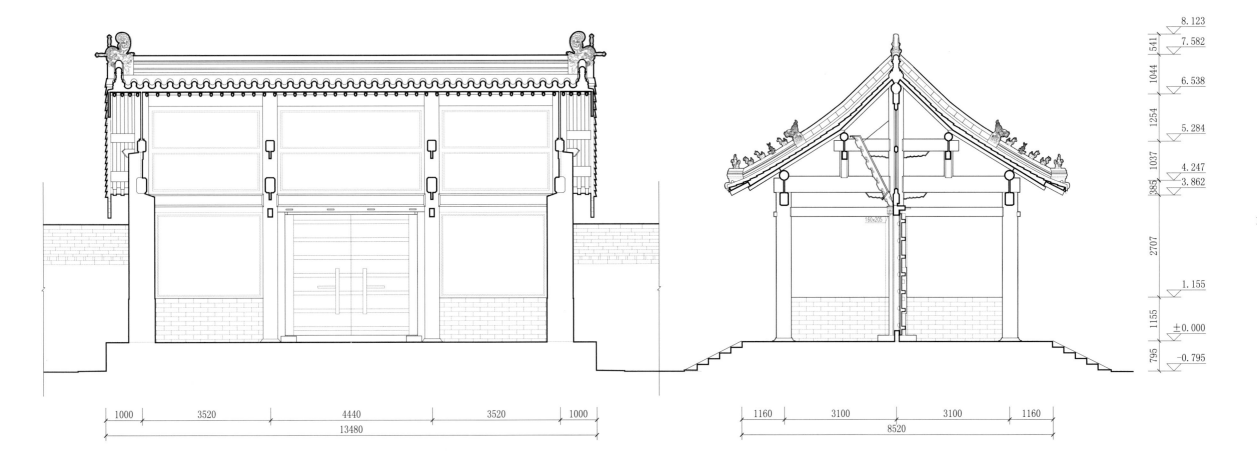

观德门纵剖面图
Longitudinal section of Guande Gate

观德门横剖面图
Cross-section of Guande Gate

正殿殿庭

The Main Court

大成门　金声门　承圣门　毓粹门

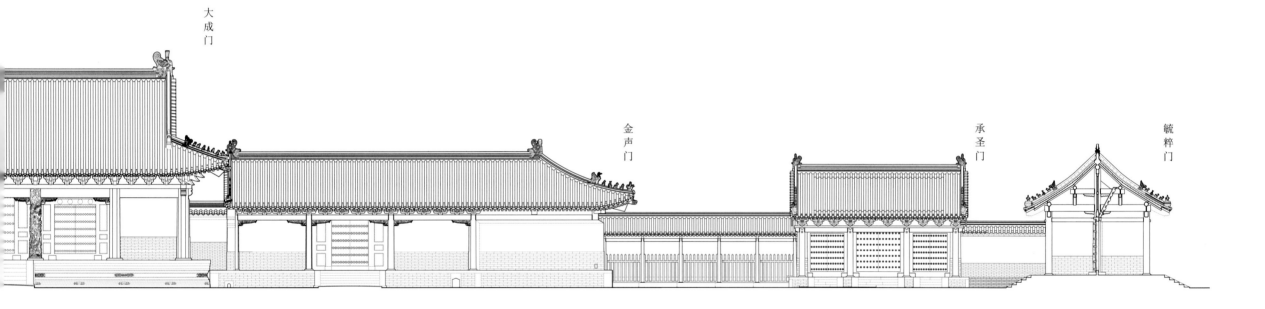

0　　3m

观德门

启圣门

玉振门

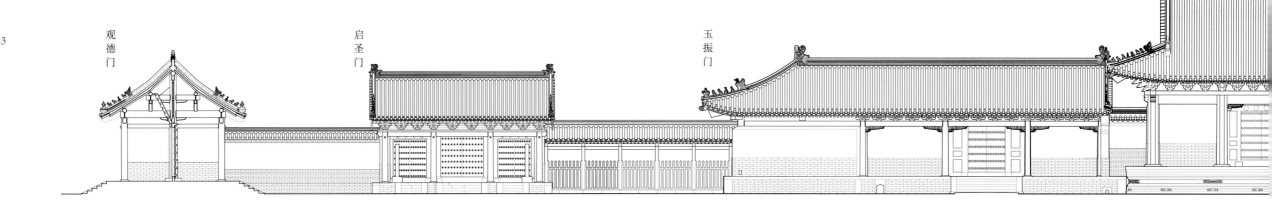

大成门组群立面图
Elevation of Dacheng Gate building complex

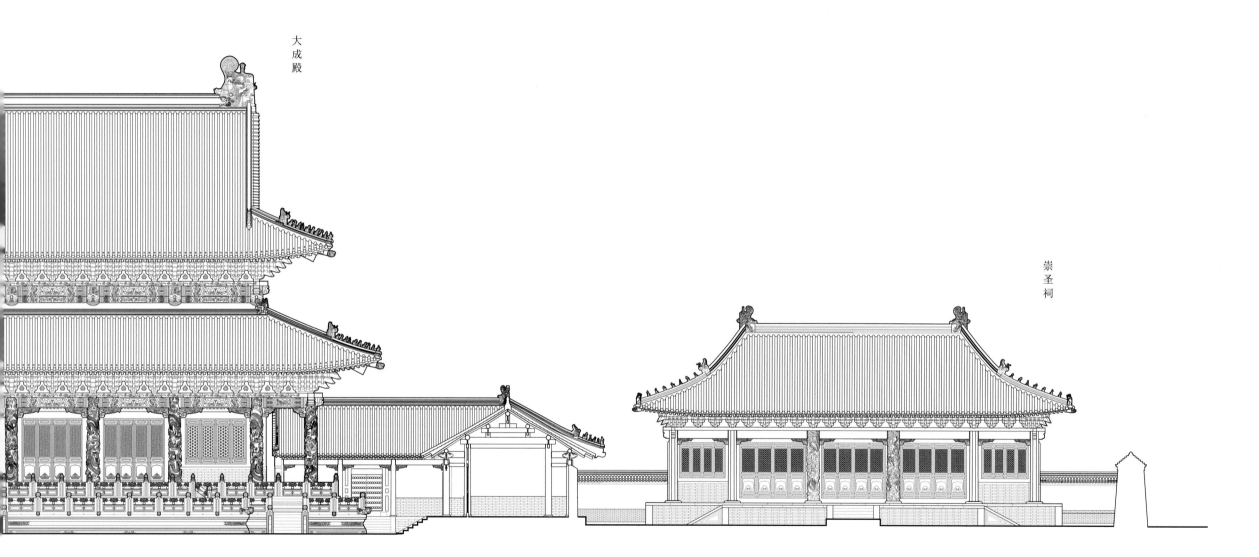

大成殿

崇圣祠

0　　3m

启圣殿

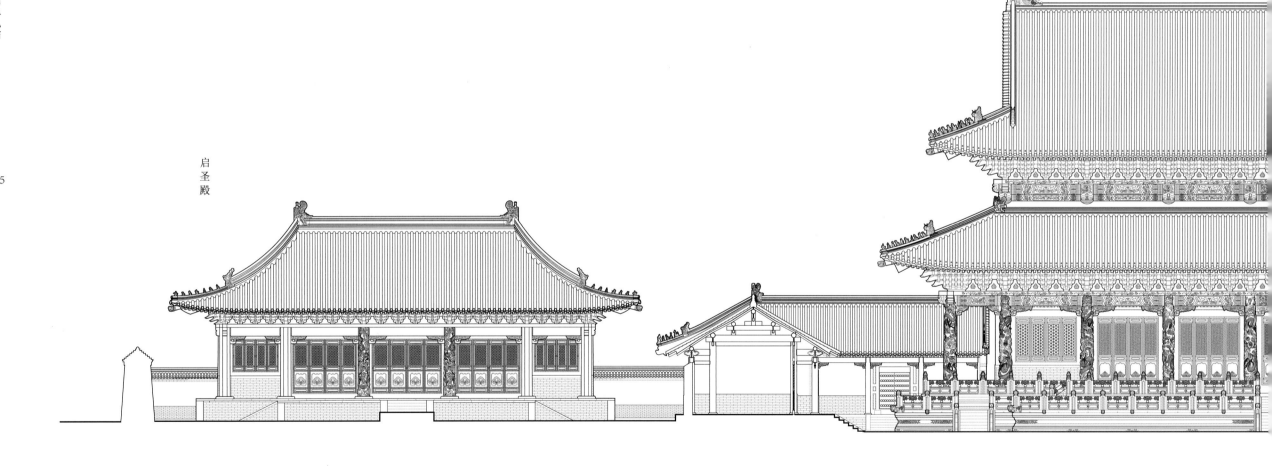

大成殿组群立面图
Elevation of Dacheng Hall building complex

寝殿

家庙

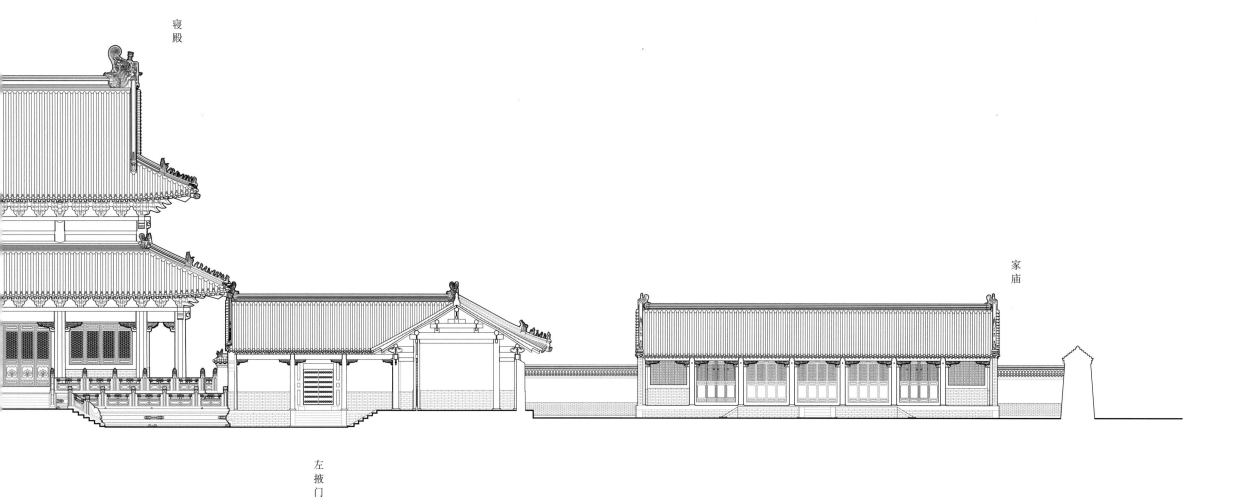

左掖门

0　　3m

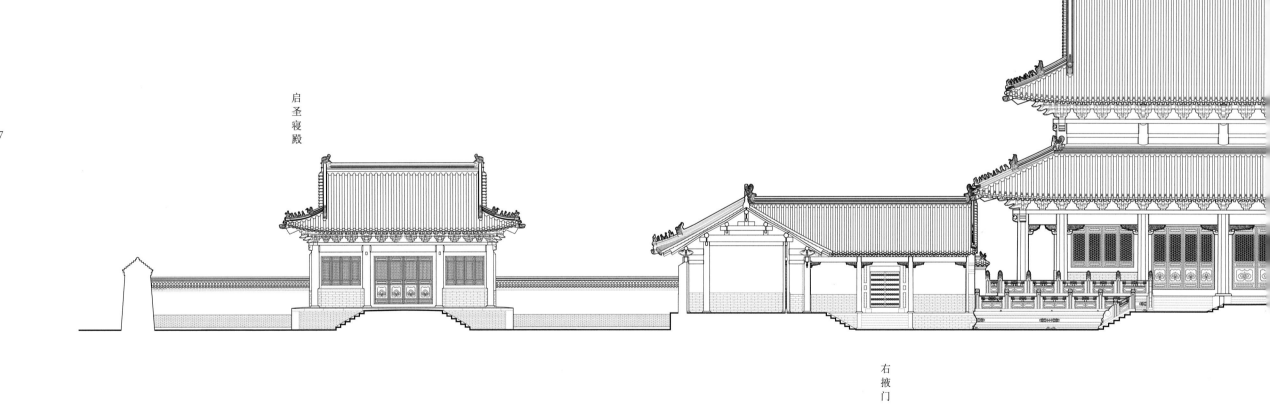

启圣寝殿

右掖门

寝殿组群立面图
Elevation of Resting Hall building complex

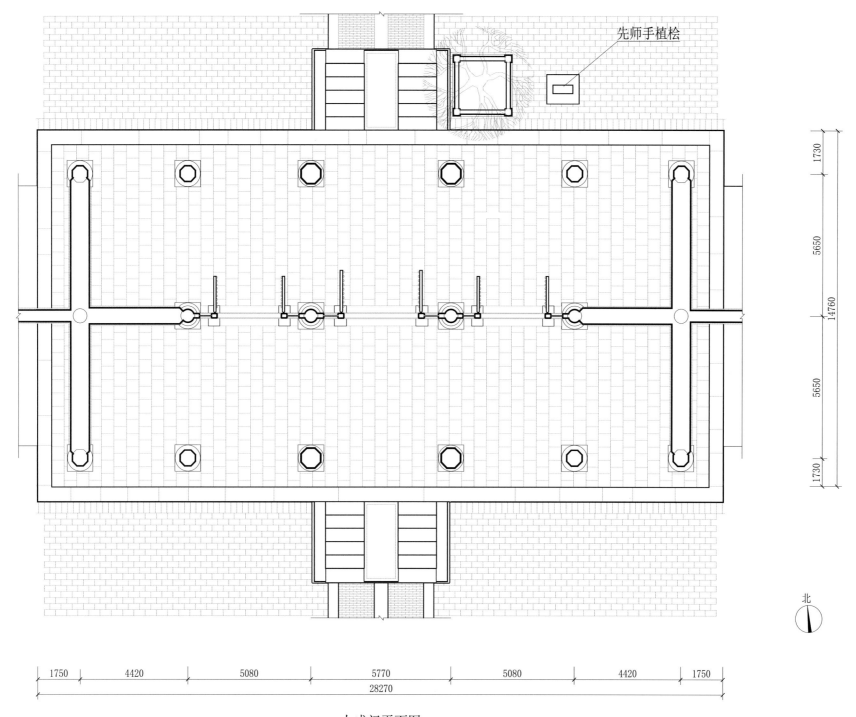

先师手植桧

1730
5650
14760
5650
1730

北

1750　4420　5080　5770　5080　4420　1750
28270

大成门平面图
Plan of Dacheng Gate

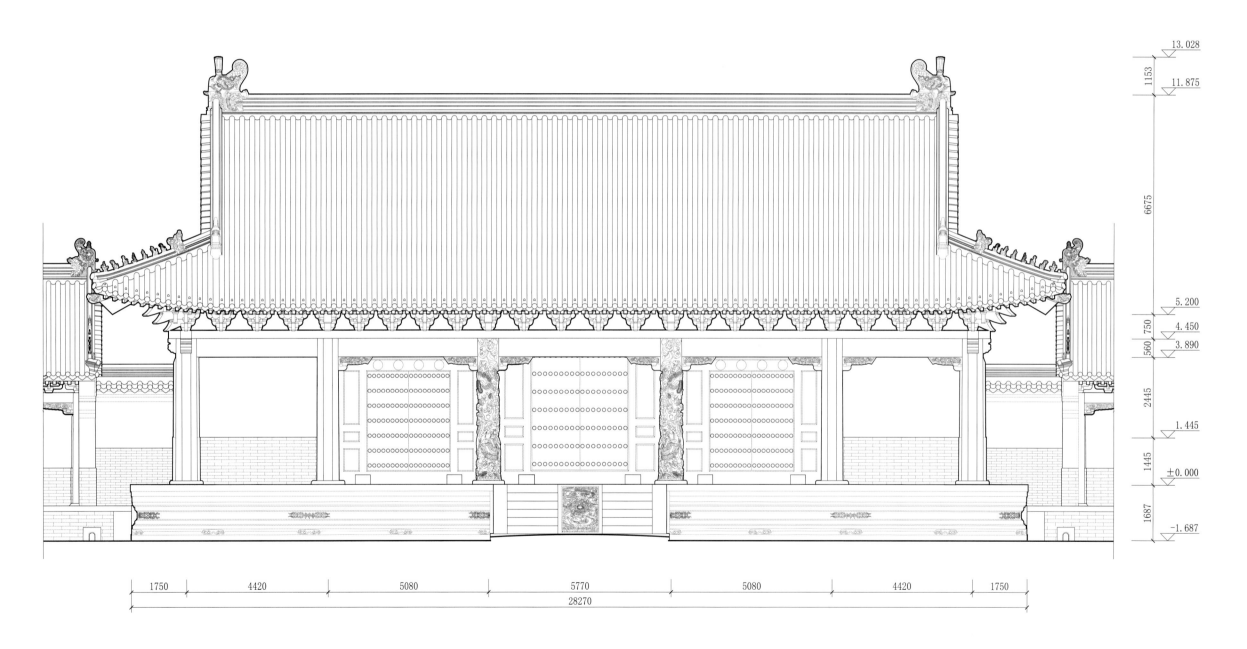

大成门正立面图
Front elevation of Dacheng Gate

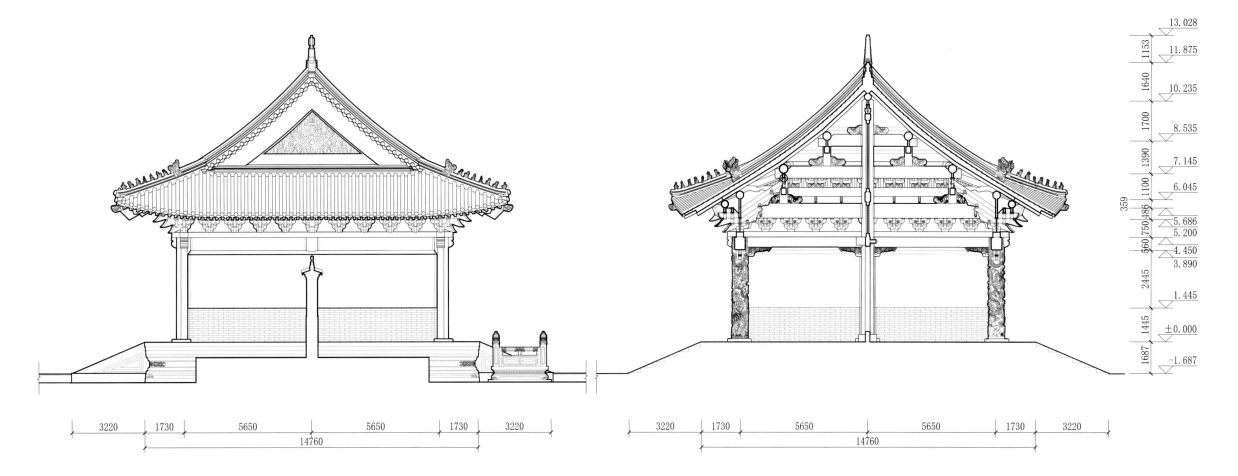

13.028

11.875

10.235

8.535

7.145

6.045

5.686

5.200

4.450

3.890

1.445

±0.000

-1.687

1153
1640
1700
1390
1100
485
750
560
2445
1445
1687

359

3220 | 1730 | 5650 | 5650 | 1730 | 3220

14760

3220 | 1730 | 5650 | 5650 | 1730 | 3220

14760

大成门侧立面图
Side elevation of Dacheng Gate

大成门横剖面图
Cross-setion of Dacheng Gate

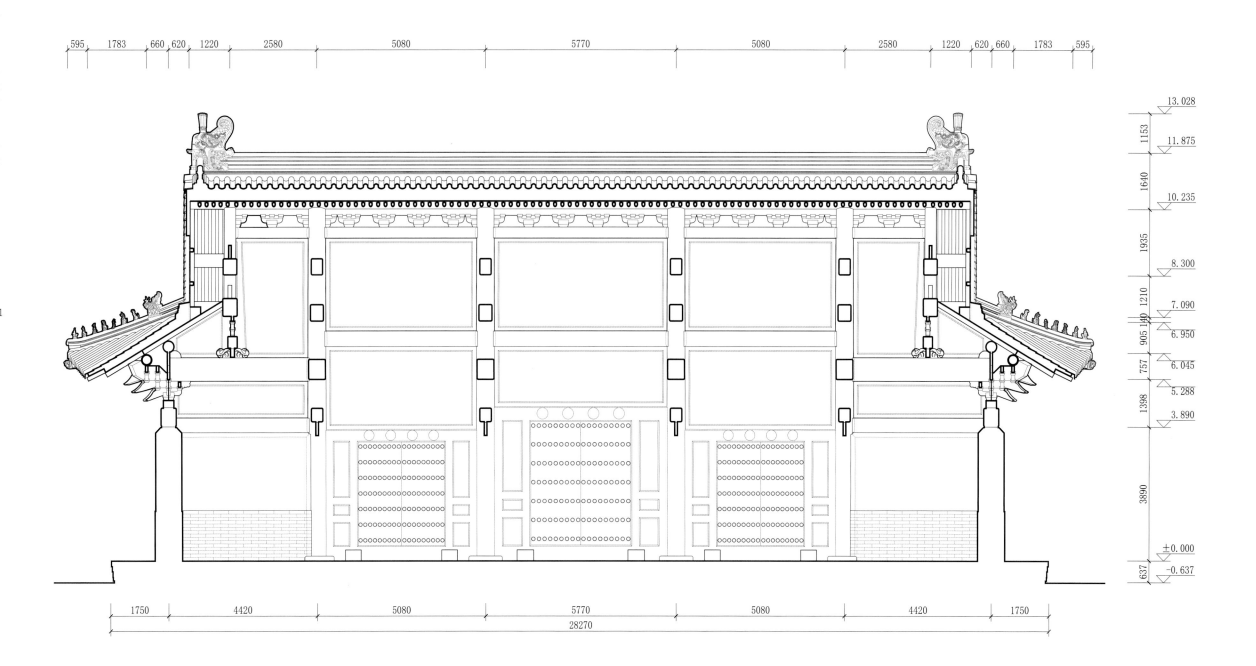

大成门纵剖面图
Longitudinal section of Dacheng Gate

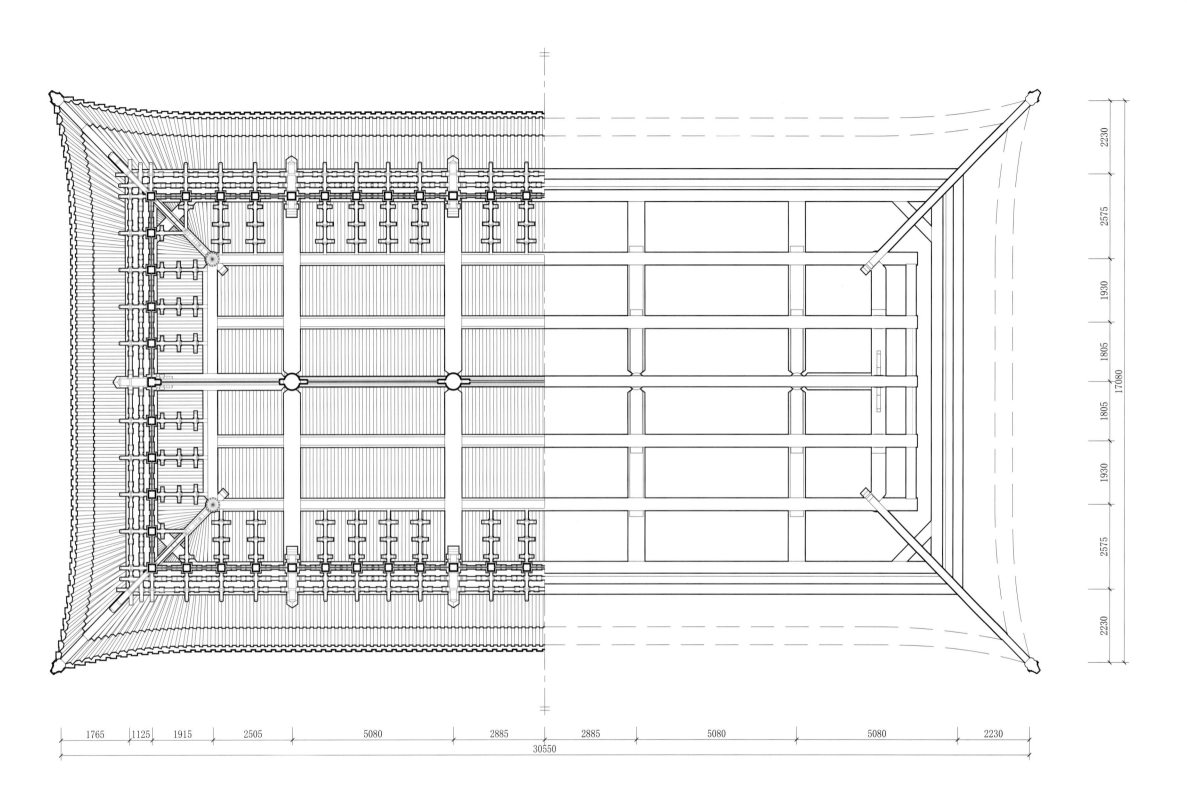

大成门梁架仰俯视图
Plan of framework of Dacheng Gate as seen from below

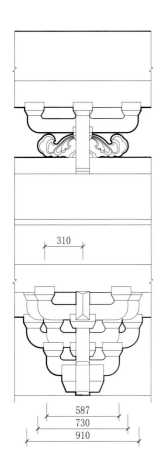

正立面图
Front elevation

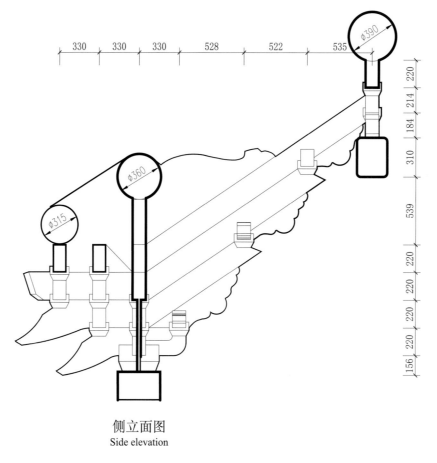

侧立面图
Side elevation

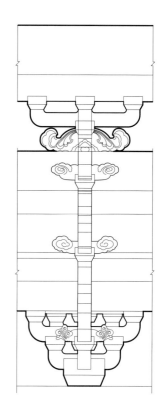

背立面图
Rear elevation

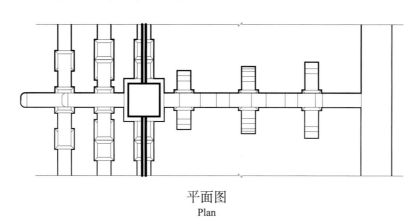

平面图
Plan

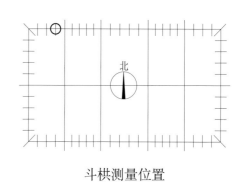

斗栱测量位置

大成门平身科斗栱大样
Intercolumnar bracket sets of Dacheng Gate

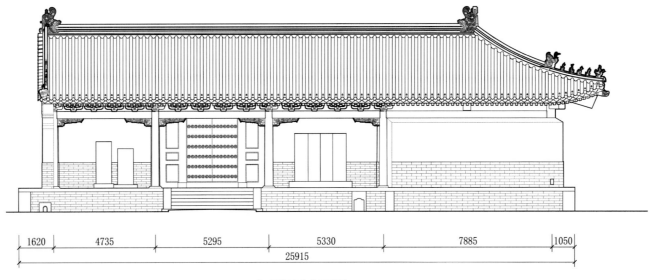

| 1620 | 4735 | 5295 | 5330 | 7885 | 1050 |
| 25915 |

金声门正立面图
Front elevation of Jinsheng Gate

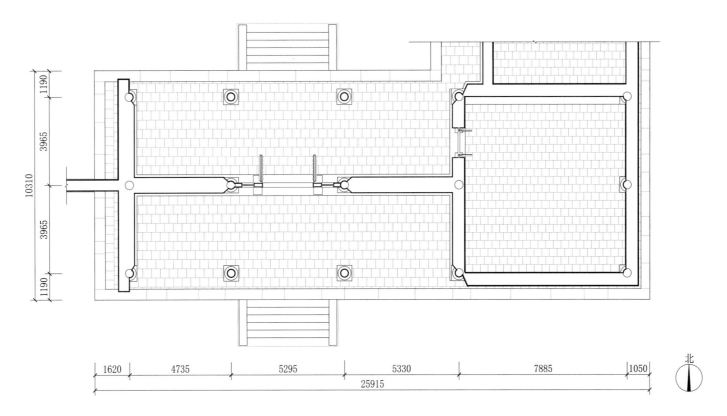

| 1620 | 4735 | 5295 | 5330 | 7885 | 1050 |
| 25915 |

北

金声门平面图
Plan of Jinsheng Gate

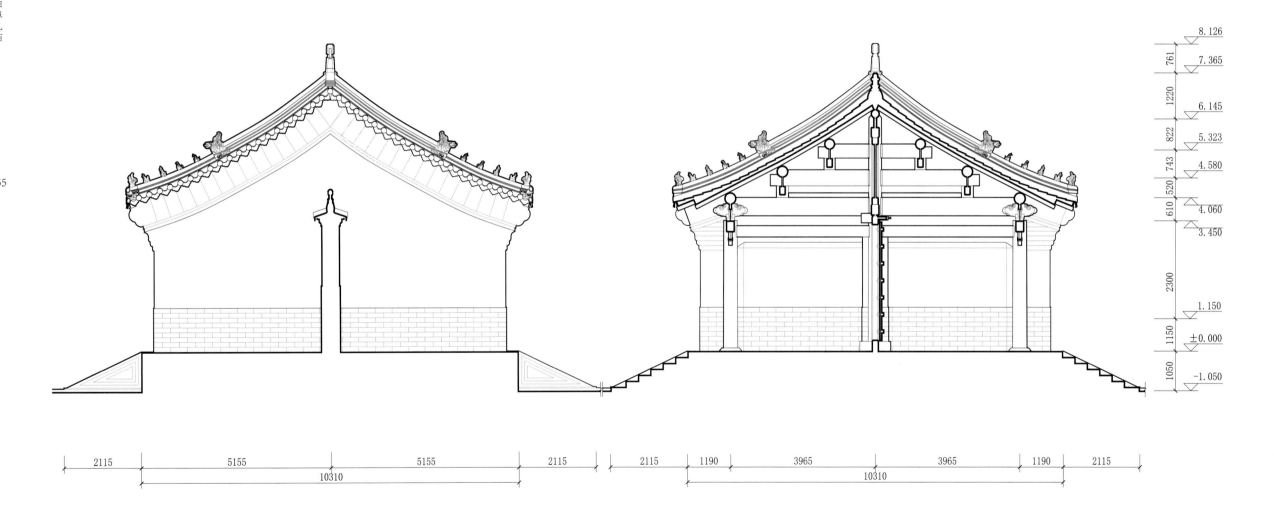

金声门侧立面图
Side elevation of Jinsheng Gate

金声门横剖面图
Cross-section of Jinsheng Gate

金声门东廊正立面图（局部）
Front elevation of east corridor of Jinsheng Gate (Local)

金声门东廊横剖面图
Cross-section of east corridor of Jinsheng Gate

杏坛
The Apricot Platform

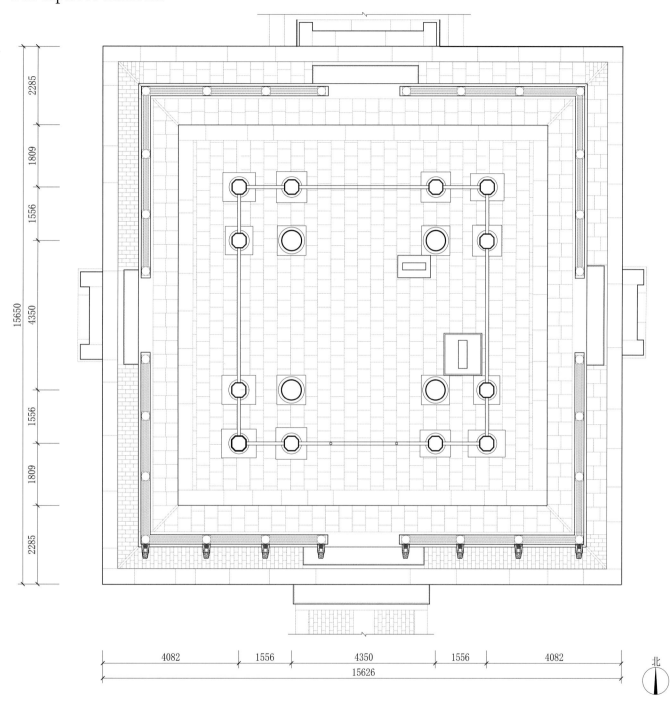

杏坛平面图
Plan of the Apricot Platform

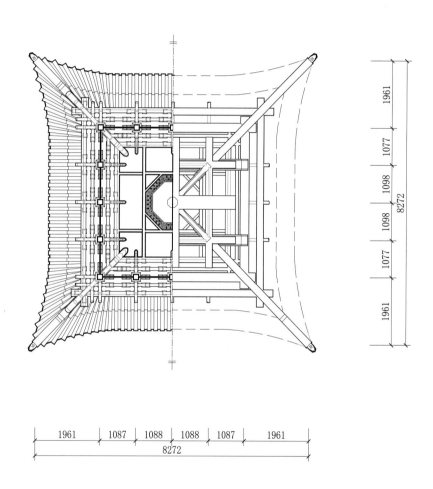

杏坛上檐梁架仰视及俯视图
Plan of upper eaves framework of the Apricot Platform, as seen from below

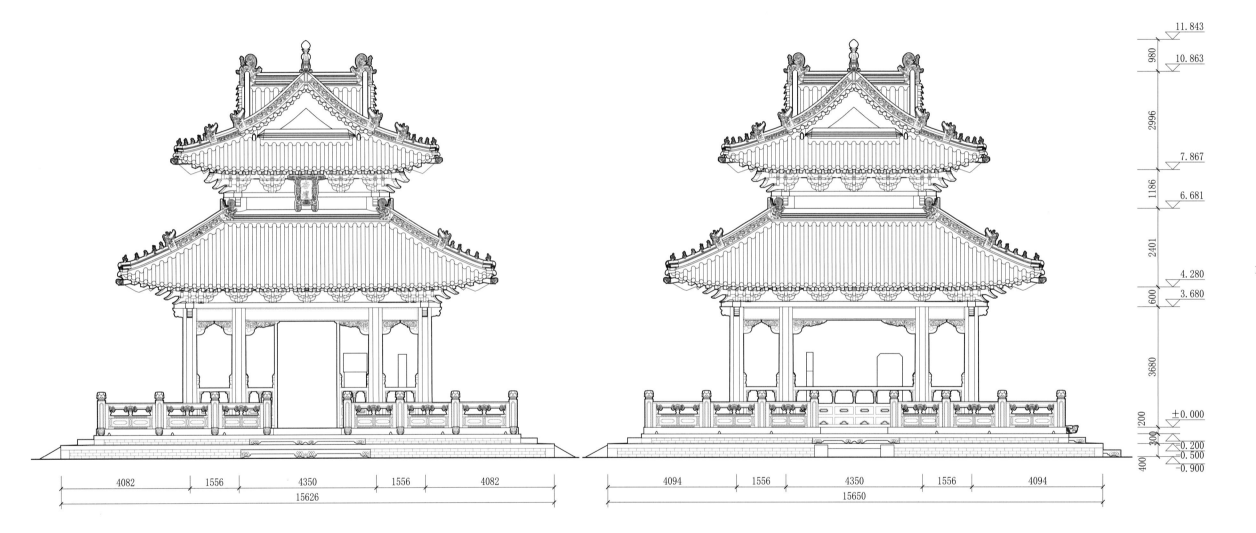

杏坛正立面图
Front elevation of the Apricot Platform

杏坛侧立面图
Side elevation of the Apricot Platform

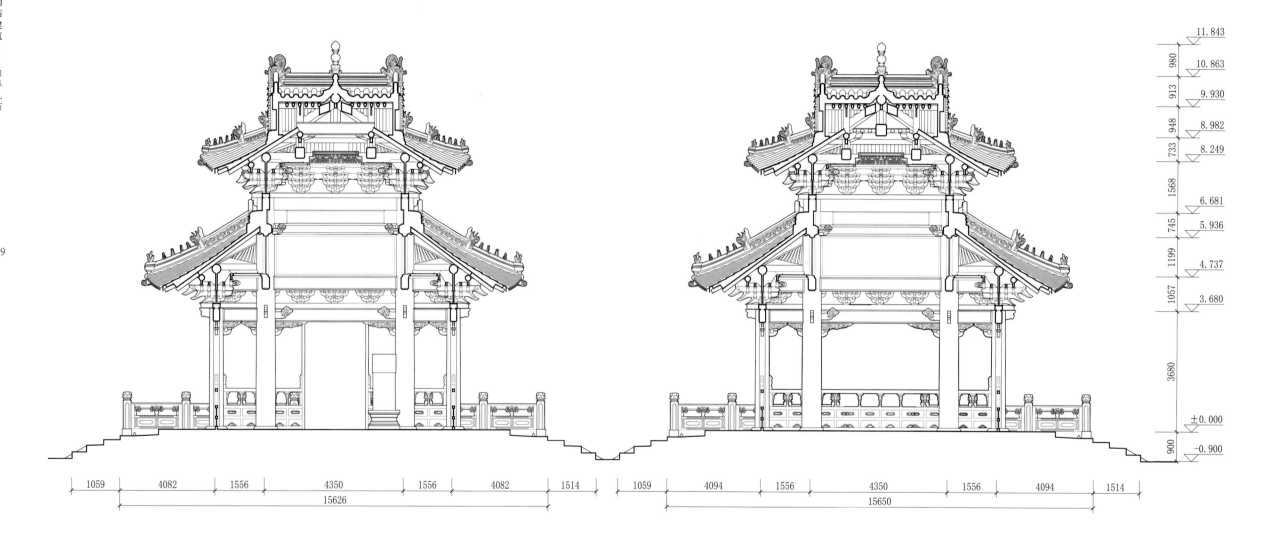

11.843
980
10.863
913
9.930
948
8.982
733
8.249
1568
6.681
745
5.936
1199
4.737
1057
3.680
3680
±0.000
900
-0.900

1059 4082 1556 4350 1556 4082 1514
15626

1059 4094 1556 4350 1556 4094 1514
15650

杏坛纵剖面图
Longitudinal section of the Apricot Platform

杏坛横剖面图
Cross-section of the Apricot Platform

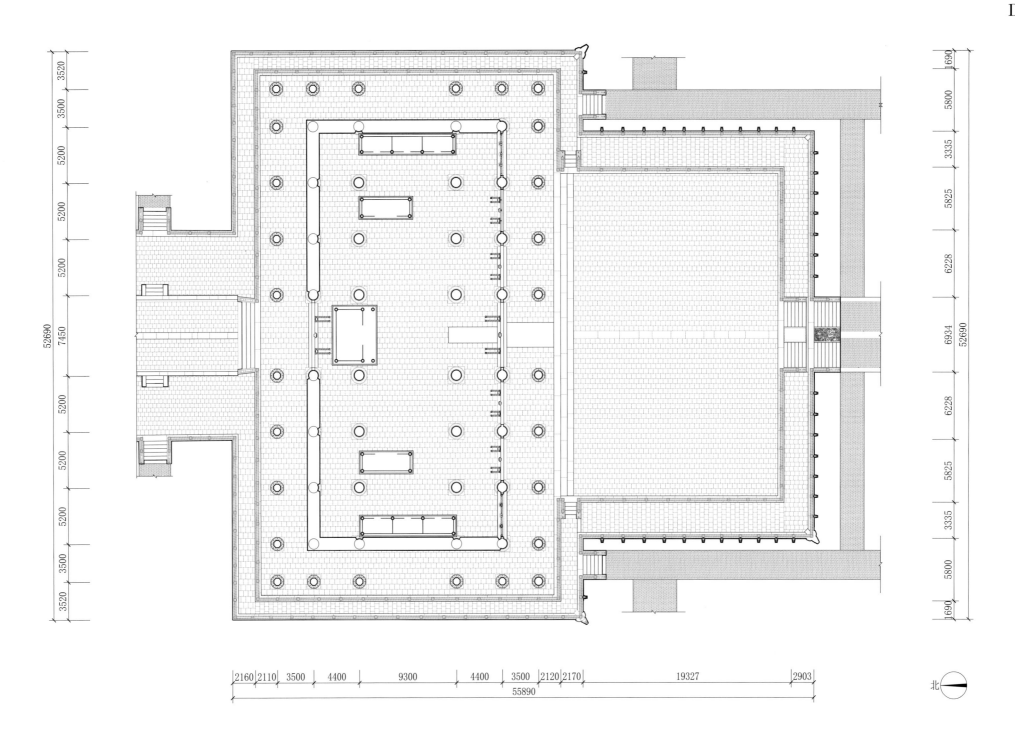

大成殿平面图
Plan of Dacheng Hall

北

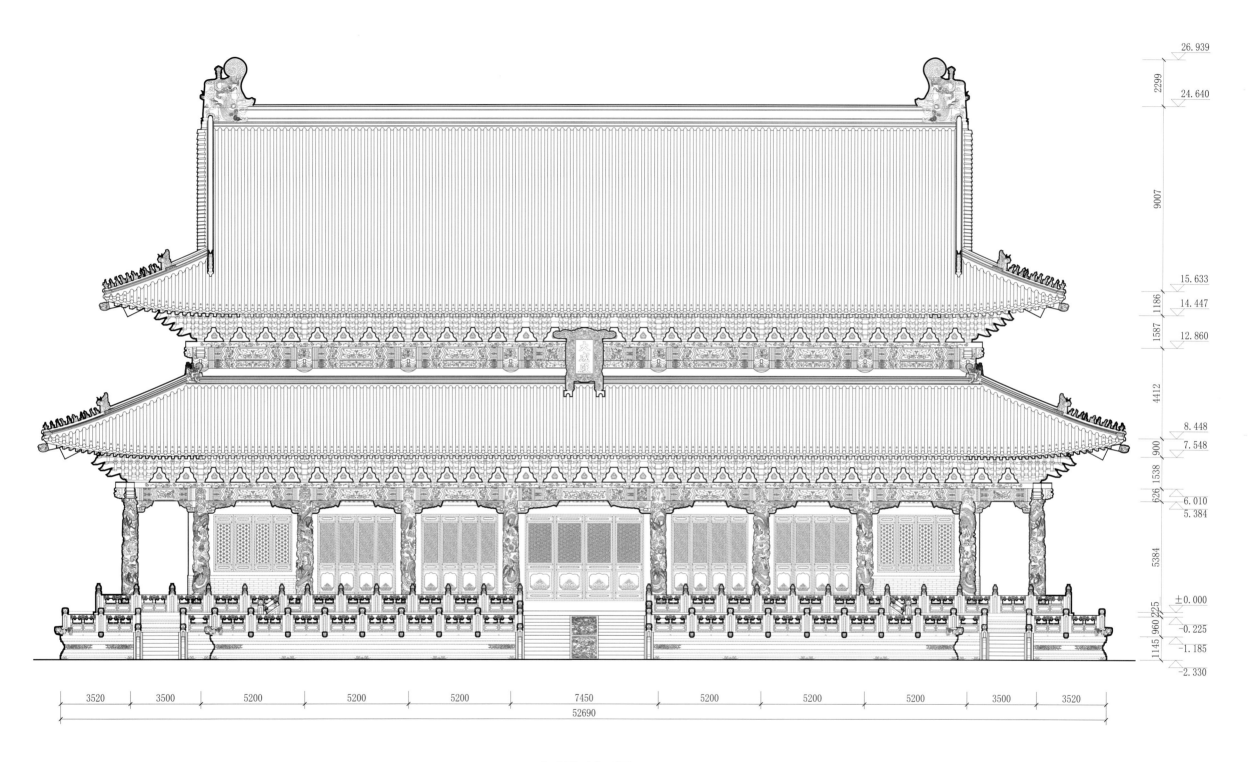

大成殿正立面图
Front elevation of Dacheng Hall

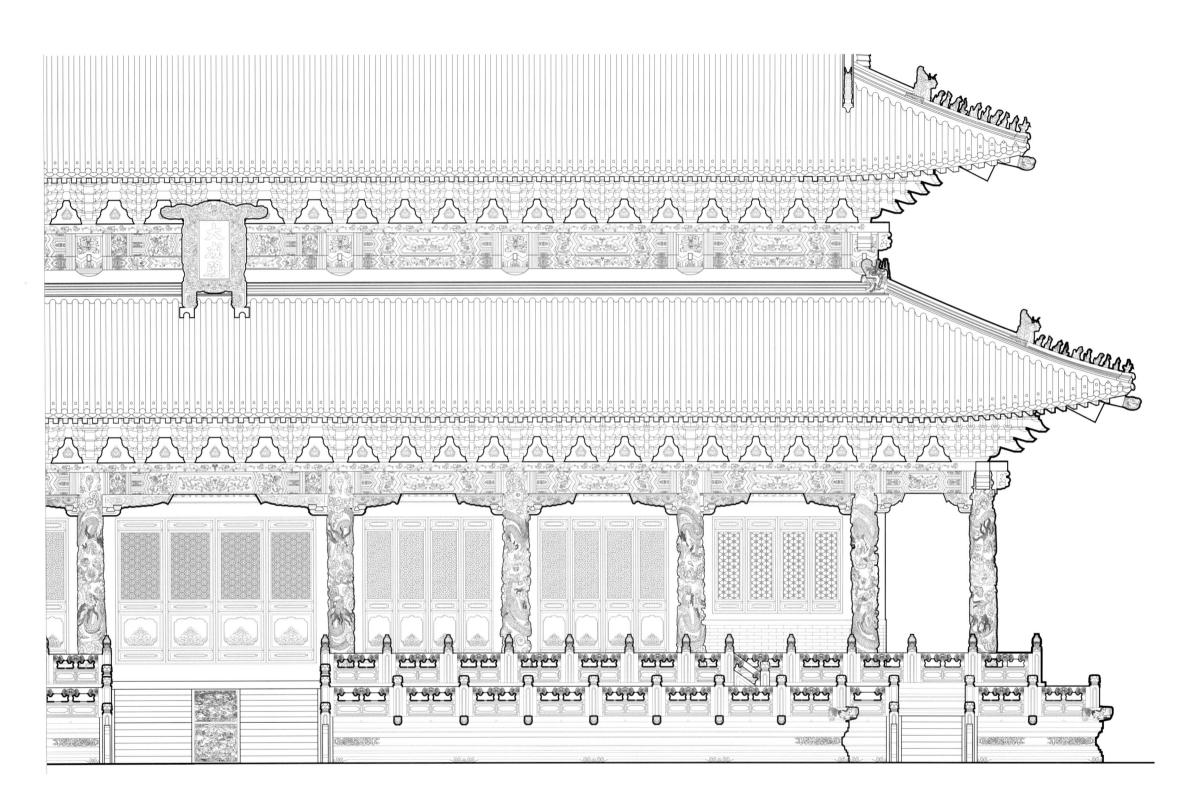

大成殿正立面图（局部）
Front elevation of Dacheng Hall(Local)

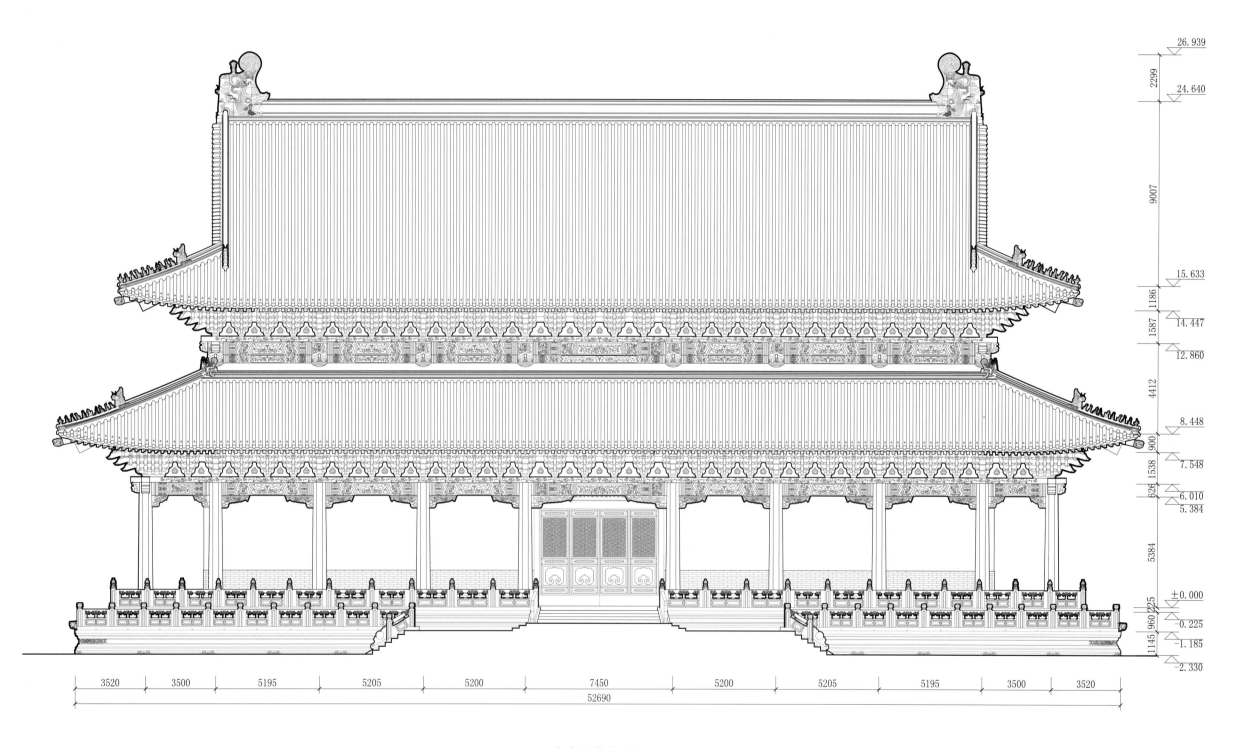

26.939

24.640

2299

9007

15.633

1186

14.447

1587

12.860

4412

8.448

900

7.548

1538

6.010

626

5.384

5384

±0.000

225

-0.225

960

-1.185

1145

-2.330

3520　3500　5195　5205　5200　7450　5200　5205　5195　3500　3520

52690

大成殿背立面图
Rear elevation of Dacheng Hall

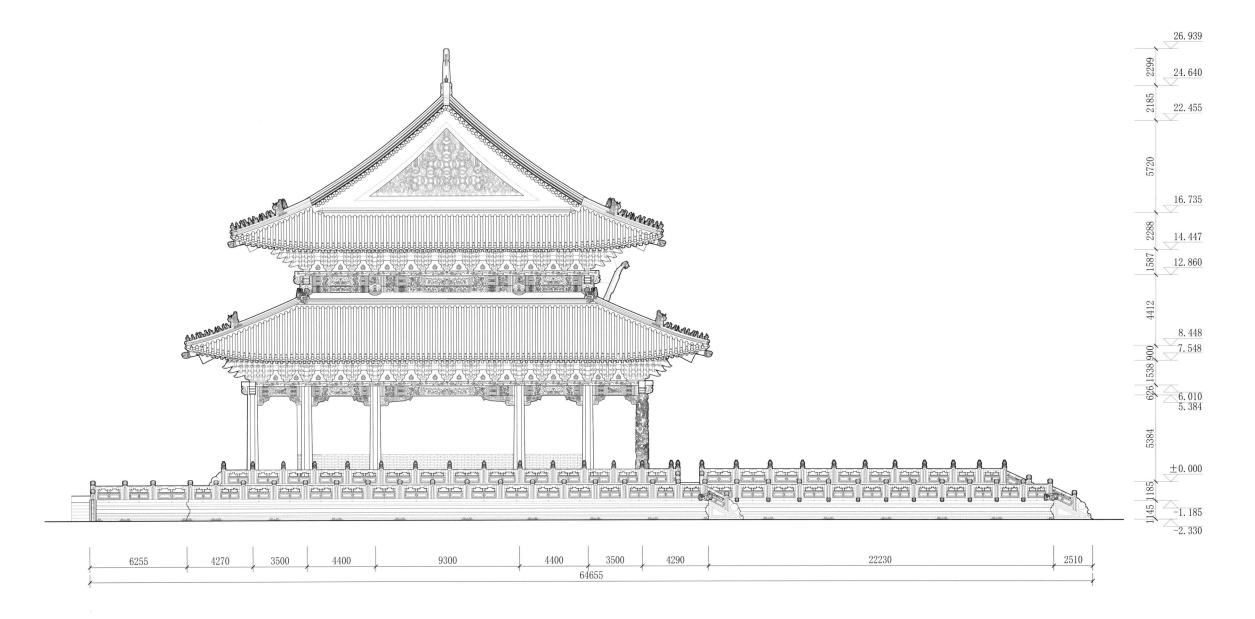

大成殿侧立面图
Side elevation of Dacheng Hall

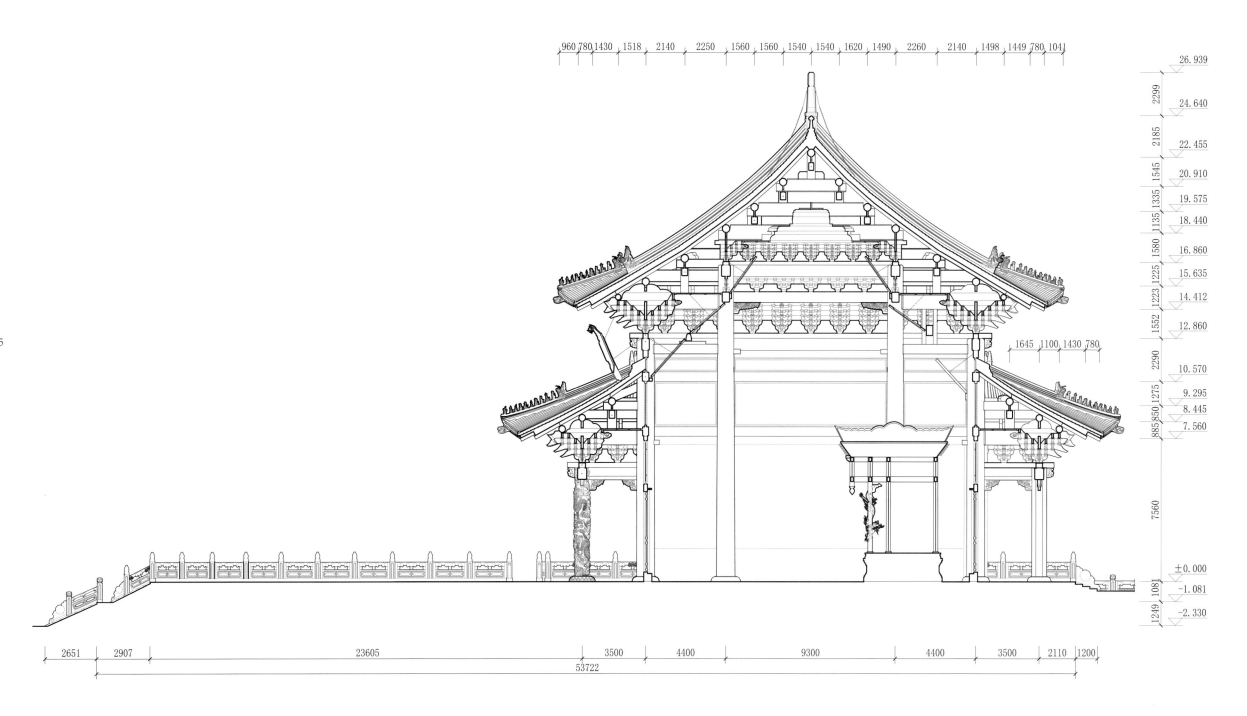

大成殿明间剖面图
Section of central bay of Dacheng Hall

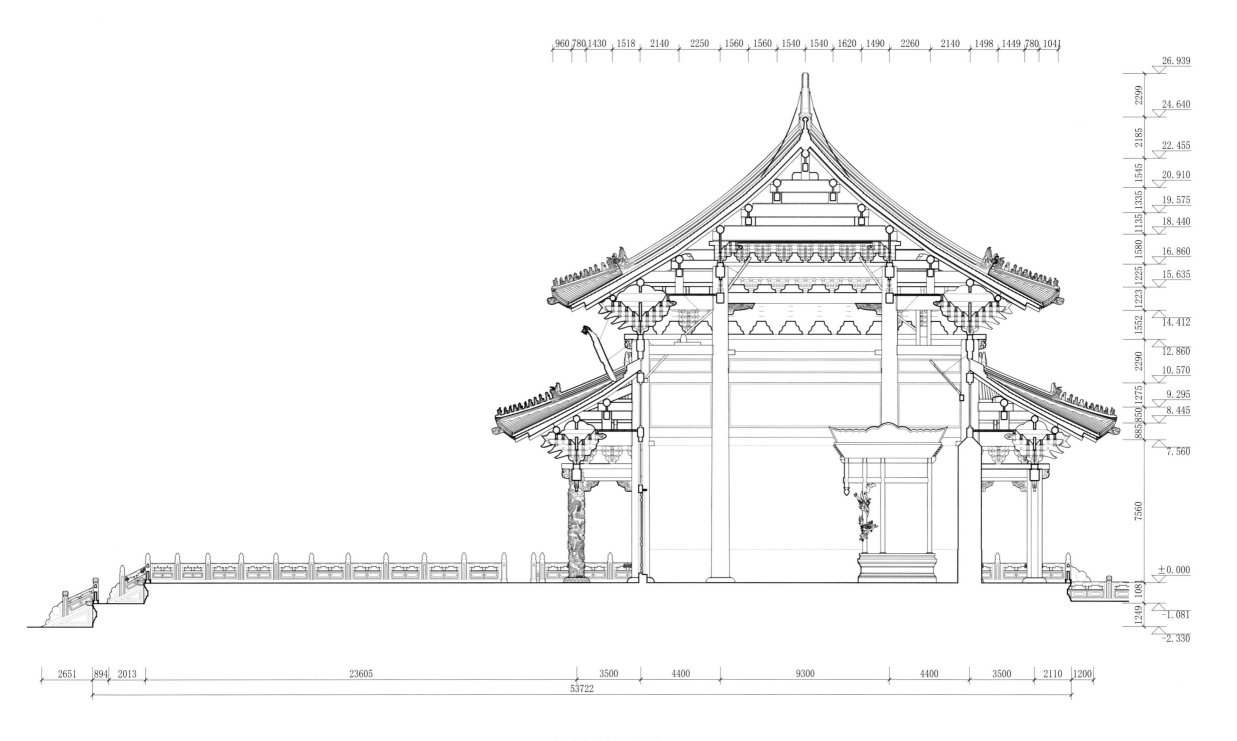

960 780 1430 1518 2140 2250 1560 1560 1540 1540 1620 1490 2260 2140 1498 1449 780 1041

26.939
2299
24.640
2185
22.455
1545
20.910
1335
19.575
1135
18.440
1580
16.860
1225
15.635
1223
14.412
1552
12.860
2290
10.570
1275
9.295
850
8.445
885
7.560
±0.000
7560
-1.081
1249 108
-2.330

2651 894 2013 23605 3500 4400 9300 4400 3500 2110 1200
53722

大成殿次间剖面图
Section of side bay of Dacheng Hall

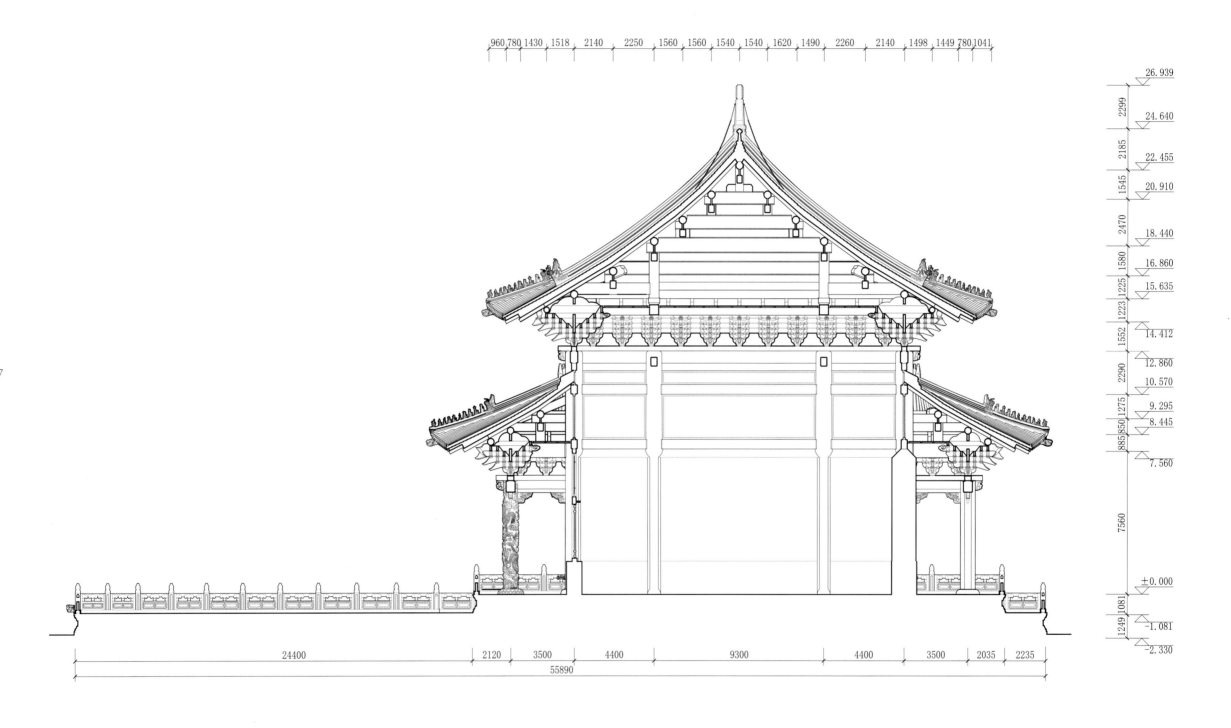

960 780 1430 1518 2140 2250 1560 1560 1540 1540 1620 1490 2260 2140 1498 1449 780 1041

26.939
2299
24.640
2185
22.455
1545
20.910
2470
18.440
1580
16.860
1225
15.635
1223
14.412
1552
12.860
2290
10.570
1275
9.295
850 885 850
8.445
7.560
7560
±0.000
1081
-1.081
1249
-2.330

24400　2120　3500　4400　9300　4400　3500　2035　2235
55890

大成殿梢间剖面图
Section of second-to-last bay of Dacheng Hall

大成殿上檐梁架仰视及俯视图
Plan of upper eaves framework of Dacheng Hall as seen from below

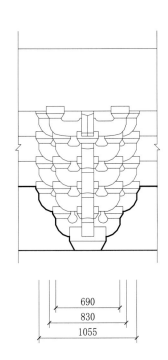

690
830
1055

正立面图
Front elevation

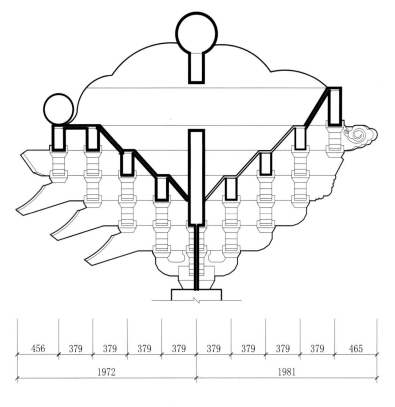

368
295
265
265
265
265
265
265
150

456 379 379 379 379 379 379 379 379 465
1972 1981

侧立面图
Side elevation

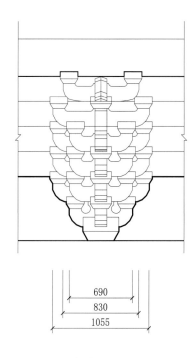

690
830
1055

背立面图
Rear elevation

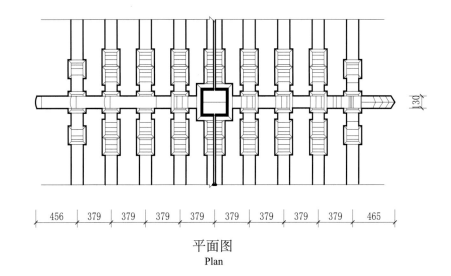

1055
830
690

130

456 379 379 379 379 379 379 379 465

平面图
Plan

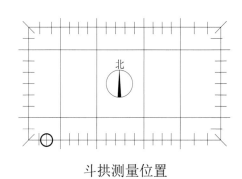

北

斗拱测量位置

大成殿上檐平身科斗栱大样图
Intercolumnar bracket sets of upper eaves of Dacheng Hall

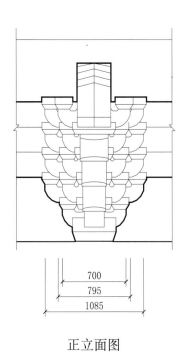

正立面图
Front elevation

侧立面图
Side elevation

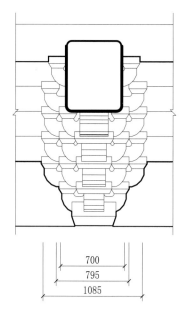

背立面图
Rear elevation

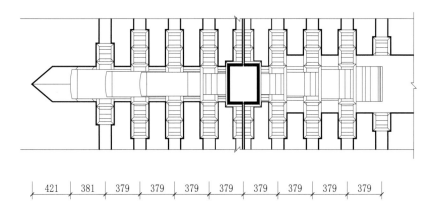

平面图
Plan

斗栱测量位置

大成殿上檐柱头科斗栱大样图
Bracket sets atop columns of upper eaves of Dacheng Hall

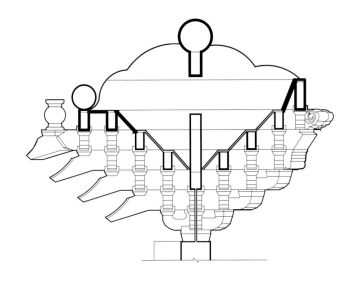

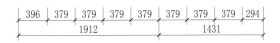

侧立面图
Side elevation

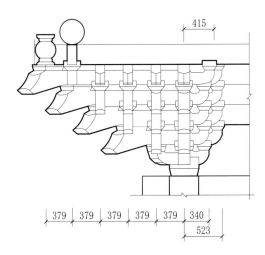

正立面图
Front elevation

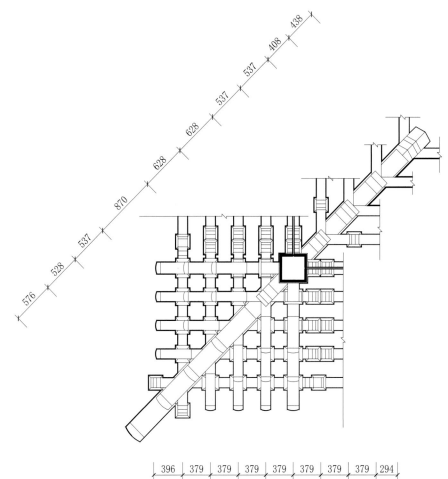

平面图
Plan

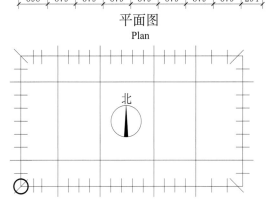

北

斗栱测量位置

大成殿上檐角科斗栱大样图
Corner bracket sets of upper eaves of Dacheng Hall

中国古建筑测绘大系·祠庙建筑——曲阜孔庙

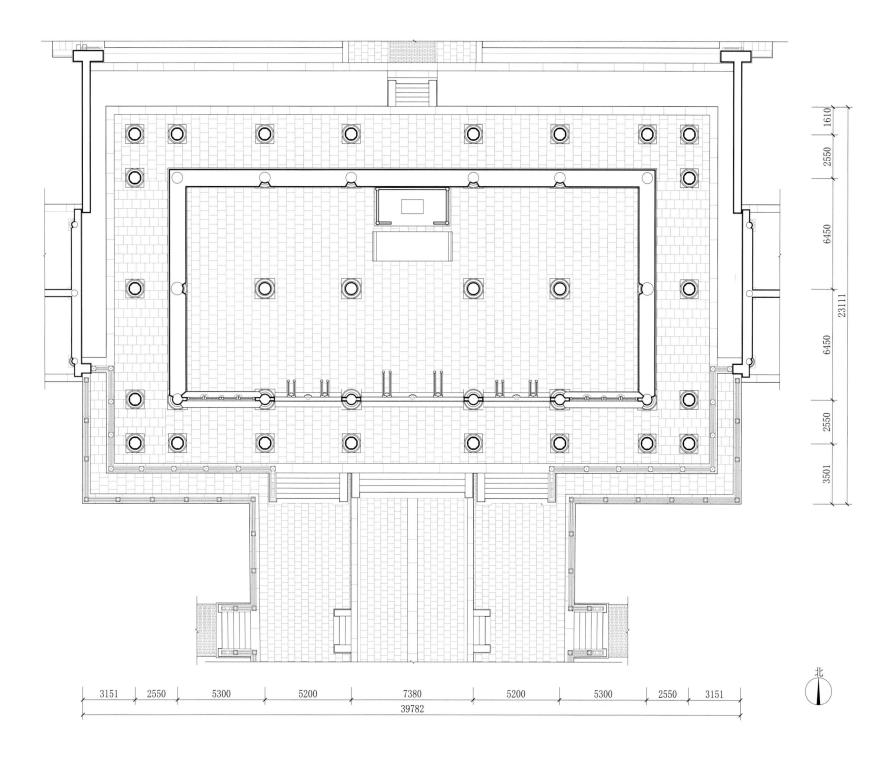

寝殿平面图
Plan of Resting Hall

北

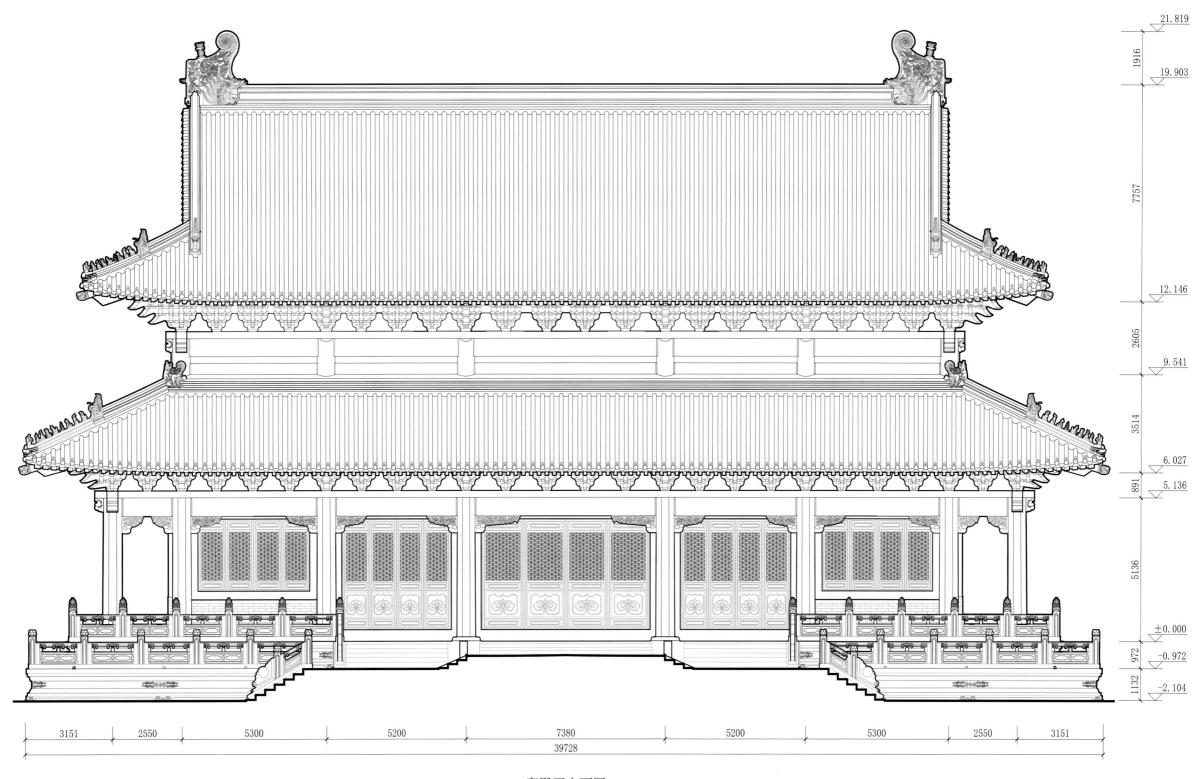

寝殿正立面图
Front elevation of resting hall

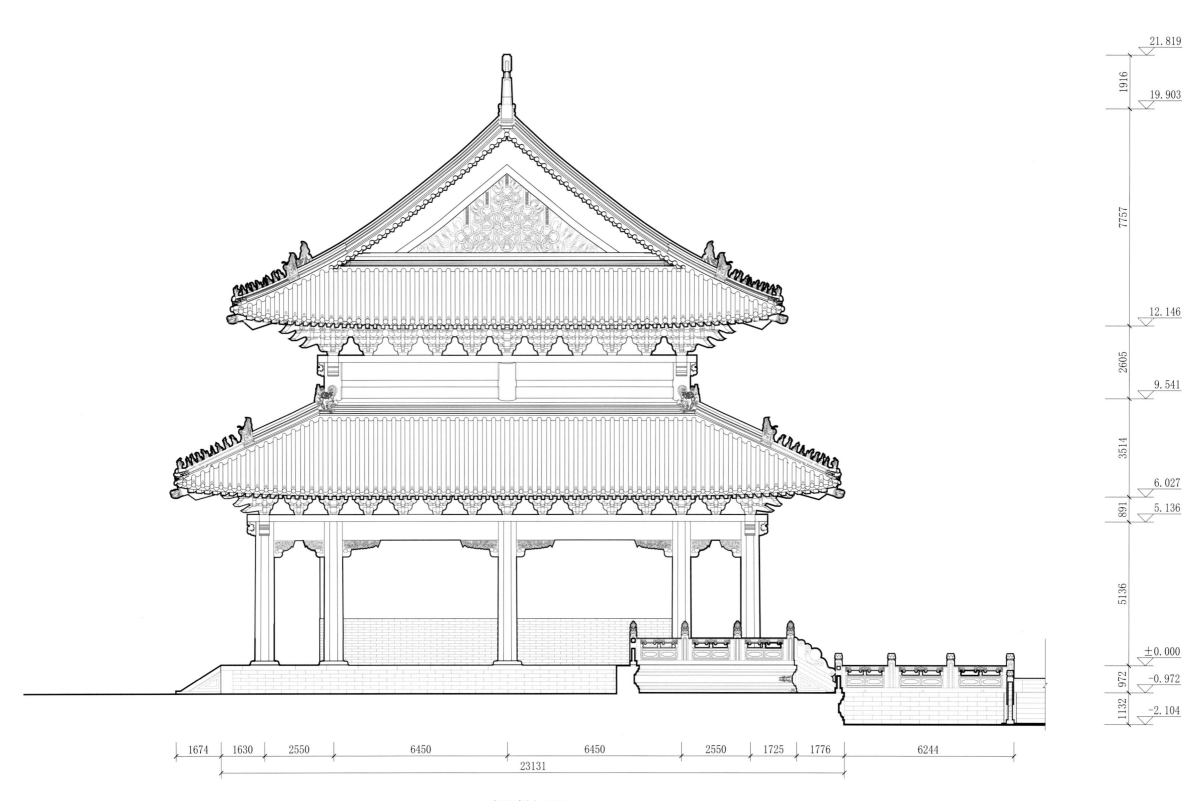

21.819

19.903

12.146

9.541

6.027

5.136

±0.000

-0.972

-2.104

1916
7757
2605
3514
891
5136
972
1132

1674 1630 2550 6450 6450 2550 1725 1776 6244
23131

寝殿侧立面图
Side elevation of resting hall

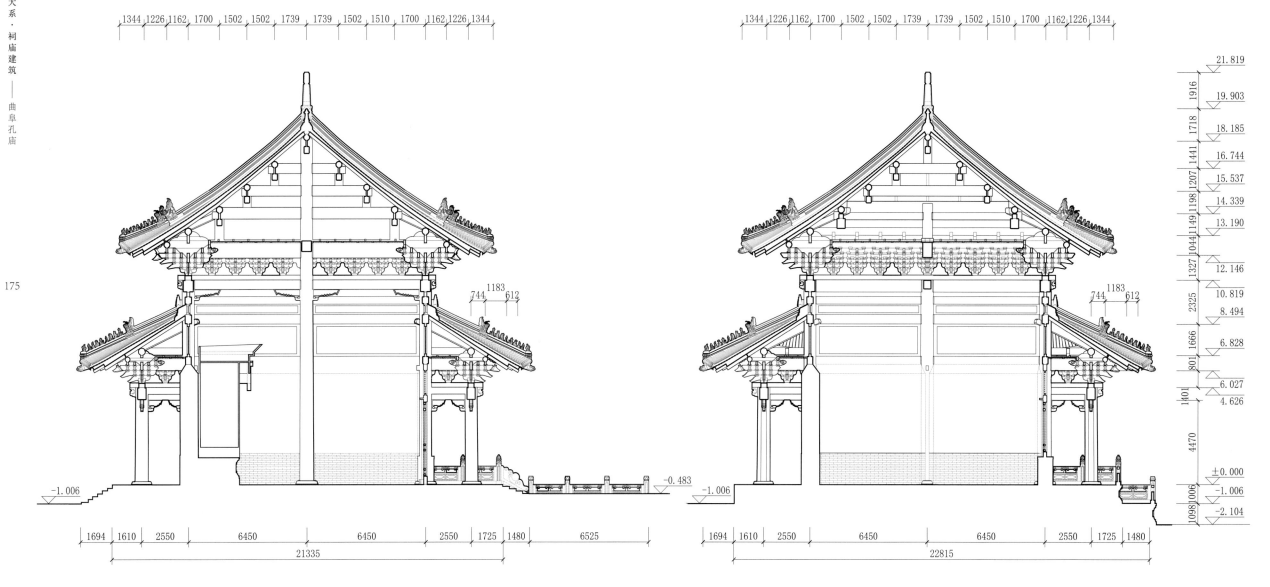

1344 1226 1162 1700 1502 1502 1739 1739 1502 1510 1700 1162 1226 1344

744 1183 612

-1.006

-0.483

1694 1610 2550 6450 6450 2550 1725 1480 6525

21335

1344 1226 1162 1700 1502 1502 1739 1739 1502 1510 1700 1162 1226 1344

21.819
1916
19.903
1718
18.185
1441
16.744
1207
15.537
1198
14.339
1149
13.190
1044
1327
12.146
2325
10.819
8.494
1666
6.828
801
6.027
1401
4.626
4470
±0.000
1006
-1.006
1098
-2.104

744 1183 612

-1.006

1694 1610 2550 6450 6450 2550 1725 1480

22815

寝殿明间剖面图
Section of central bay of resting hall

寝殿梢间剖面图
Section of second-to-last bay of resting hall

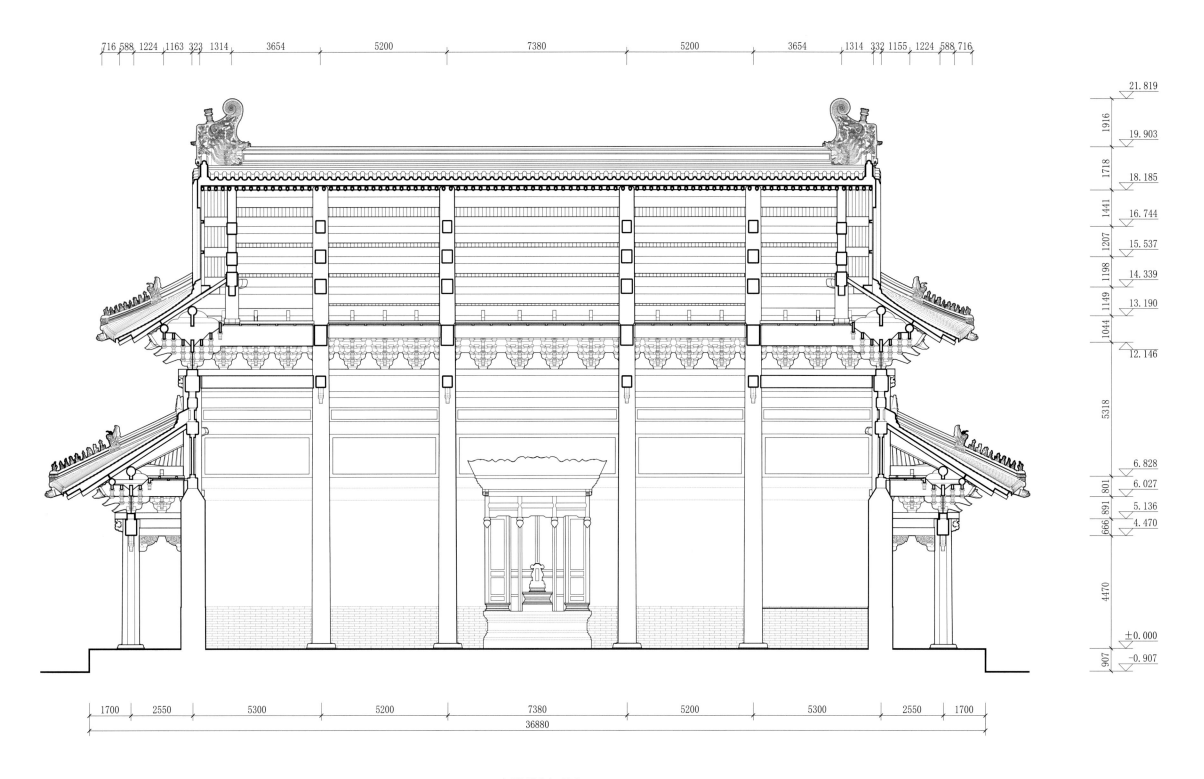

寝殿纵剖面图
Longitudinal section of resting hall

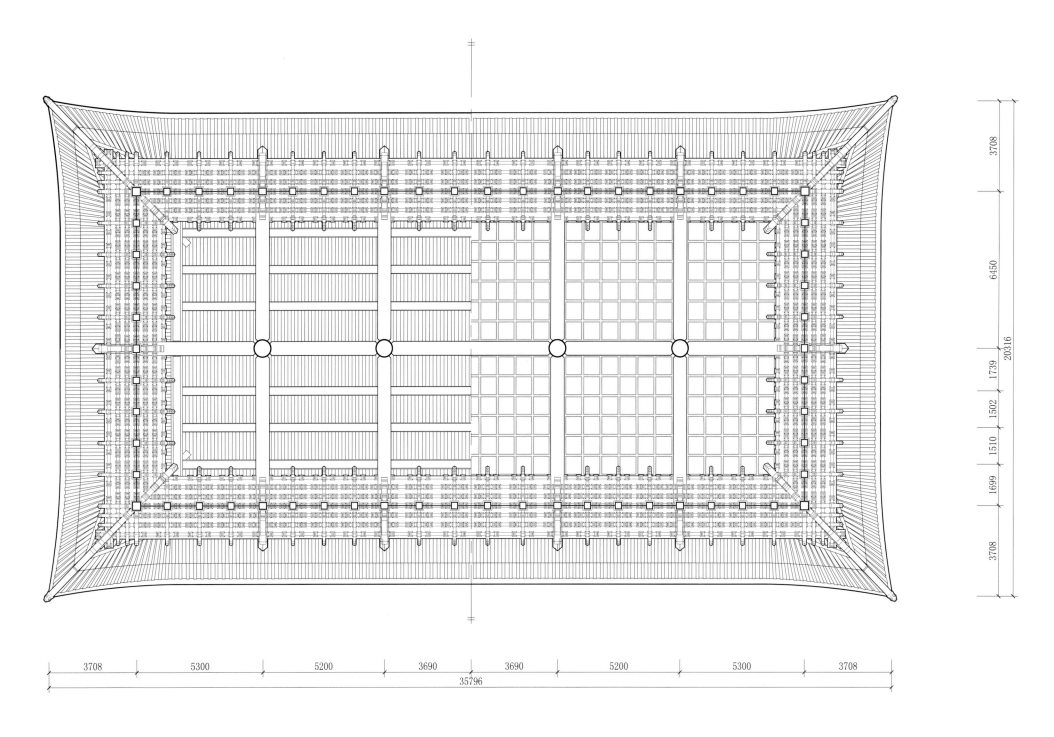

寝殿上檐梁架仰视俯视图
Plan of upper eaves framework of resting hall as seen from below

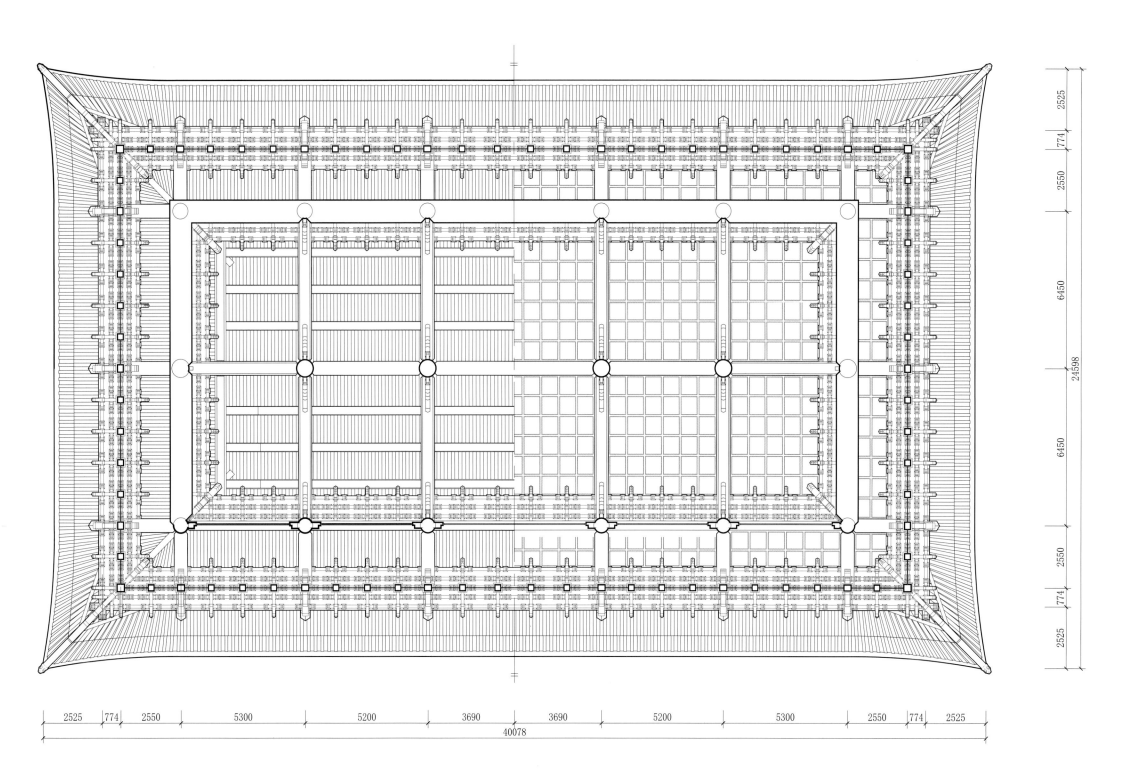

寝殿下檐梁架仰视俯视图

Plan of lower eaves framework of resting hall as seen from below

殿庭东院

The East Court

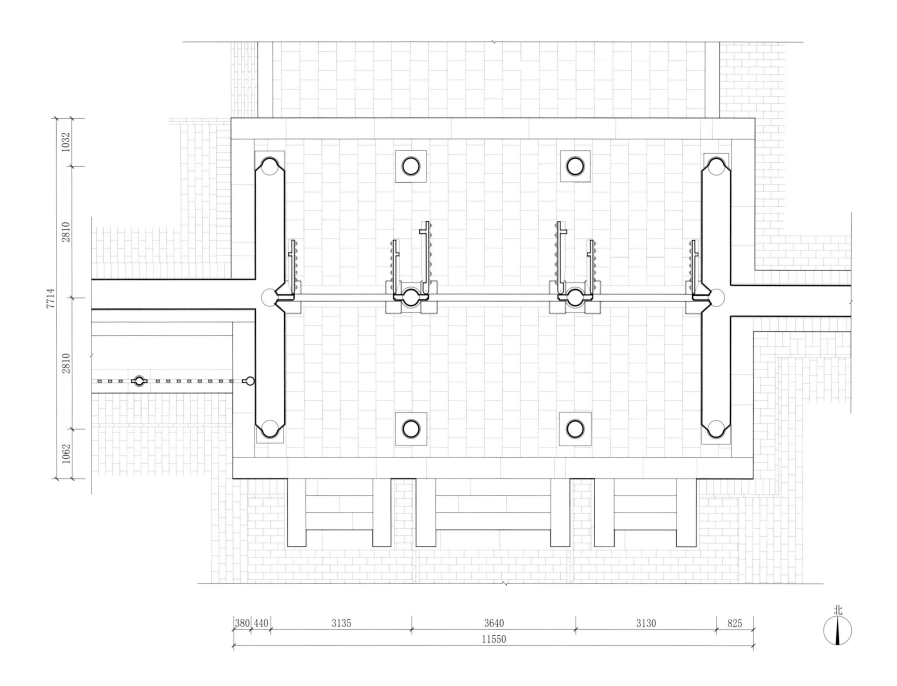

承圣门平面图
Plan of Chengsheng Gate

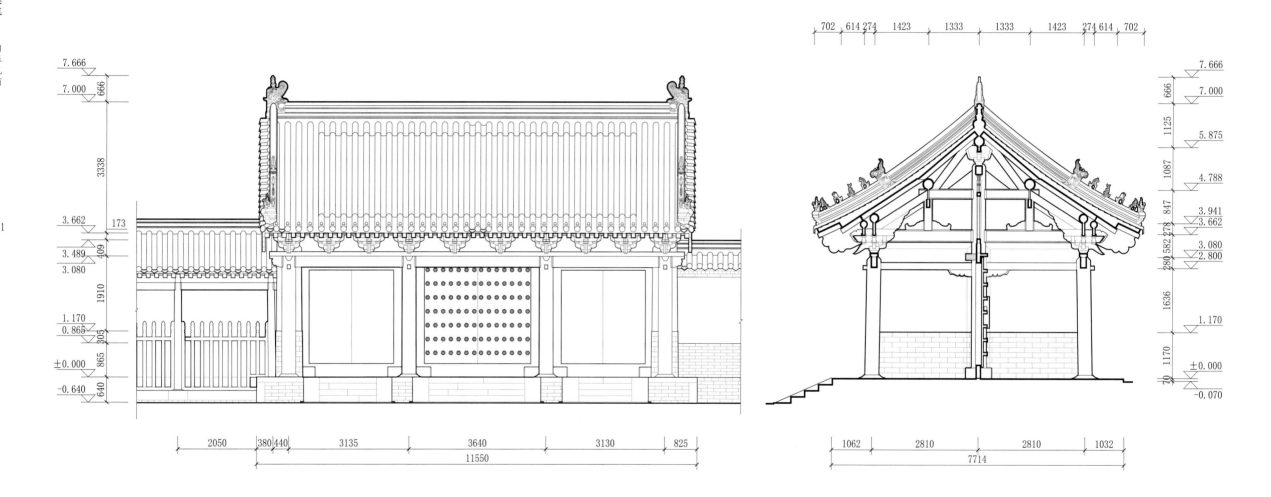

承圣门正立面图
Front elevation of Chengsheng Gate

承圣门明间剖面图
Section of central bay of Chengsheng Gate

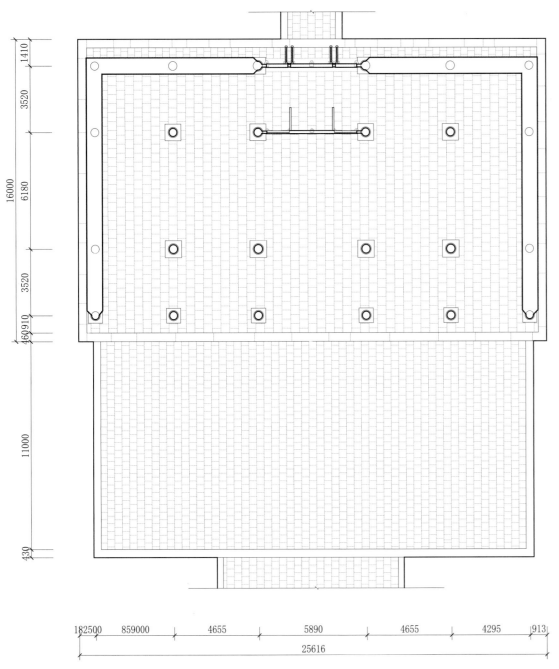

诗礼堂平面图
Plan of Shili Hall

诗礼堂梁架仰视图
Plan of framework of Shili Hall as seen from below

北

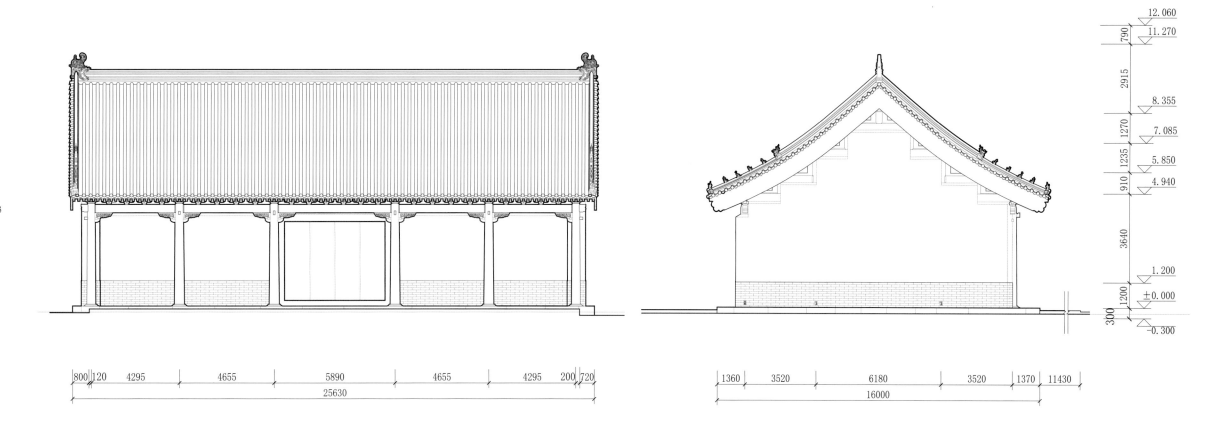

诗礼堂正立面图
Front elevation of Shili Hall

诗礼堂侧立面图
Side elevation of Shili Hall

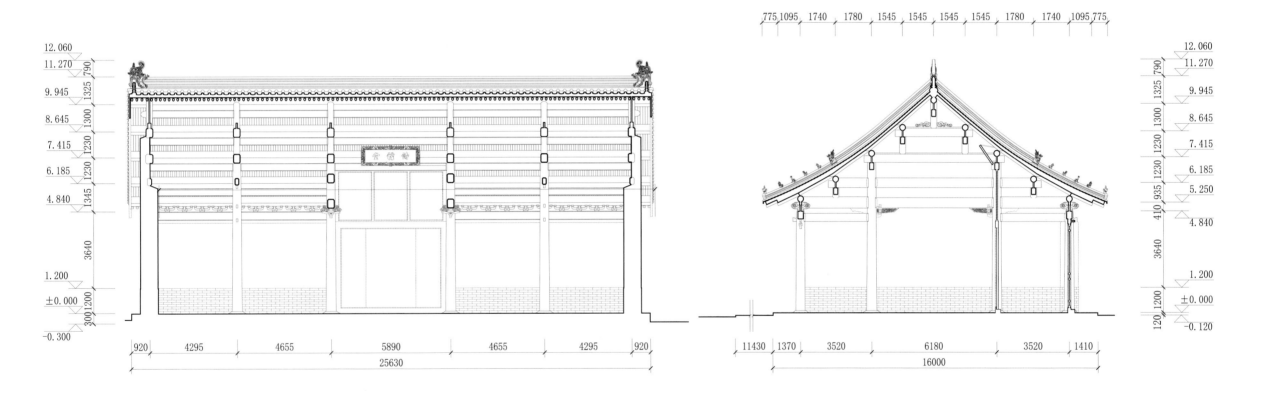

诗礼堂纵剖面图
Longitudinal section of Shili Hall

诗礼堂明间剖面图
Section of central bay of Shili Hall

礼器库
Sacrificial Vessel Storeroom

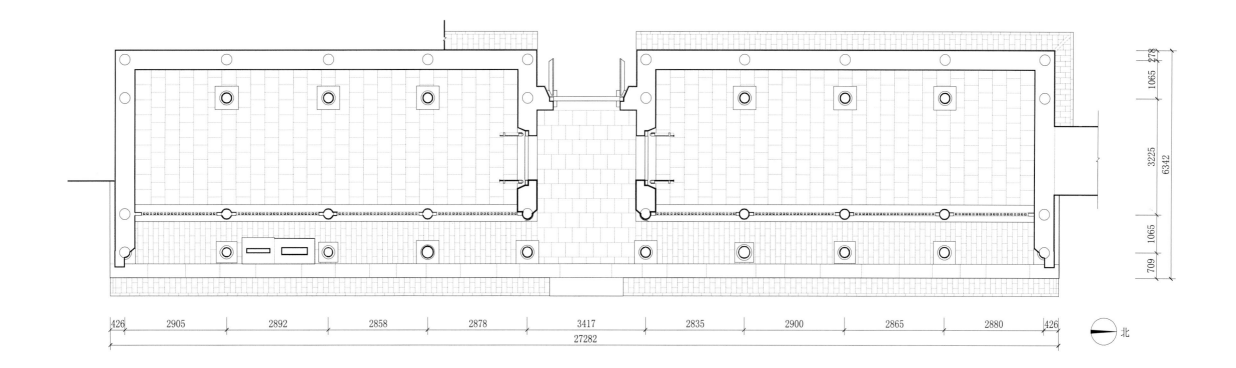

| 426 | 2905 | 2892 | 2858 | 2878 | 3417 | 2835 | 2900 | 2865 | 2880 | 426 |

27282

北

礼器库平面图
Plan of sacrificial vessel storeroom

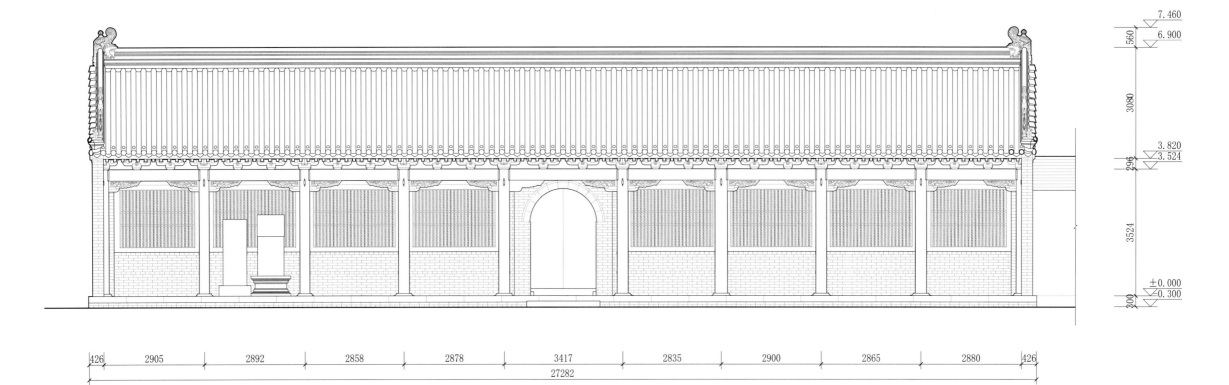

礼器库正立面图
Front elevation of sacrificial vessel storeroom

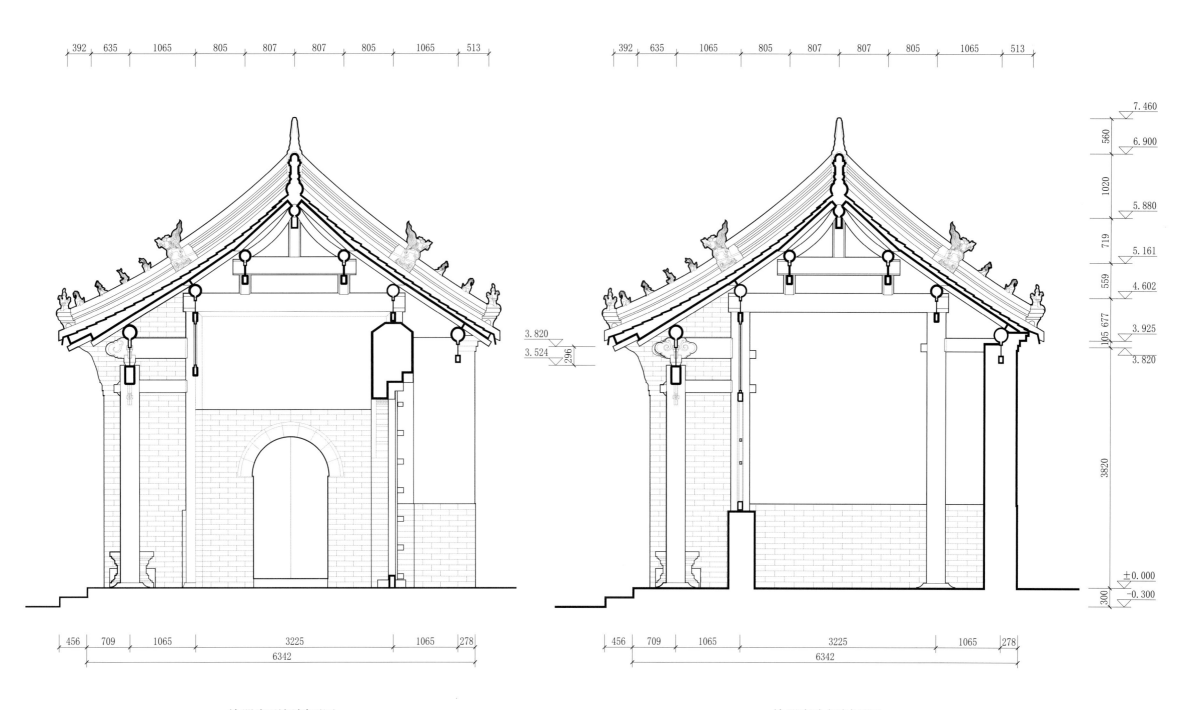

礼器库明间剖面图
Section of central bay of sacrificial vessel storeroom

礼器库次间剖面图
Section of side bay of sacrificial vessel storeroom

中国古建筑测绘大系·祠庙建筑——曲阜孔庙

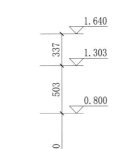

孔宅故井南立面图
South elevation of the well at the old residence

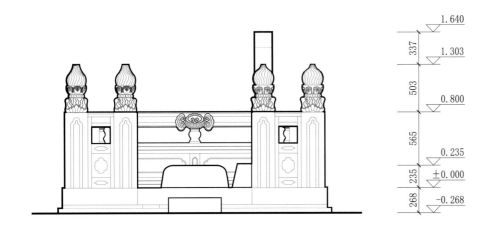

孔宅故井东立面图
East elevation of the well at the old residence

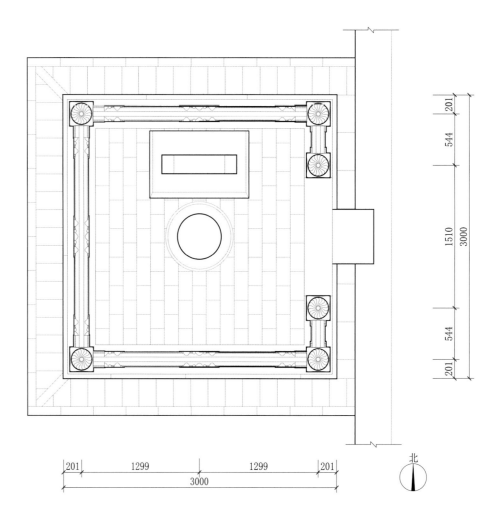

孔宅故井平面图
Plan of the well at the old residence

北

故井碑亭
Stele pavilion of the well at the old residence

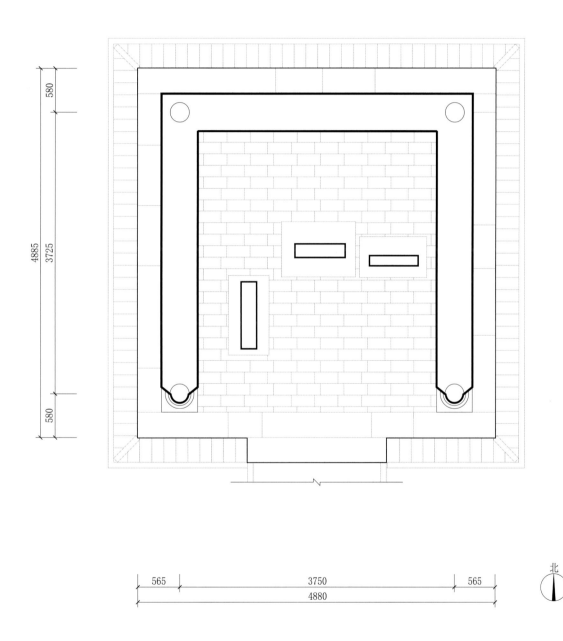

北

故井碑亭平面图
Plan of stele pavilion of the well at the old residence

故井碑亭梁架仰俯视图
Plan of framework of stele pavilion of the well at the old residence, as seen from below

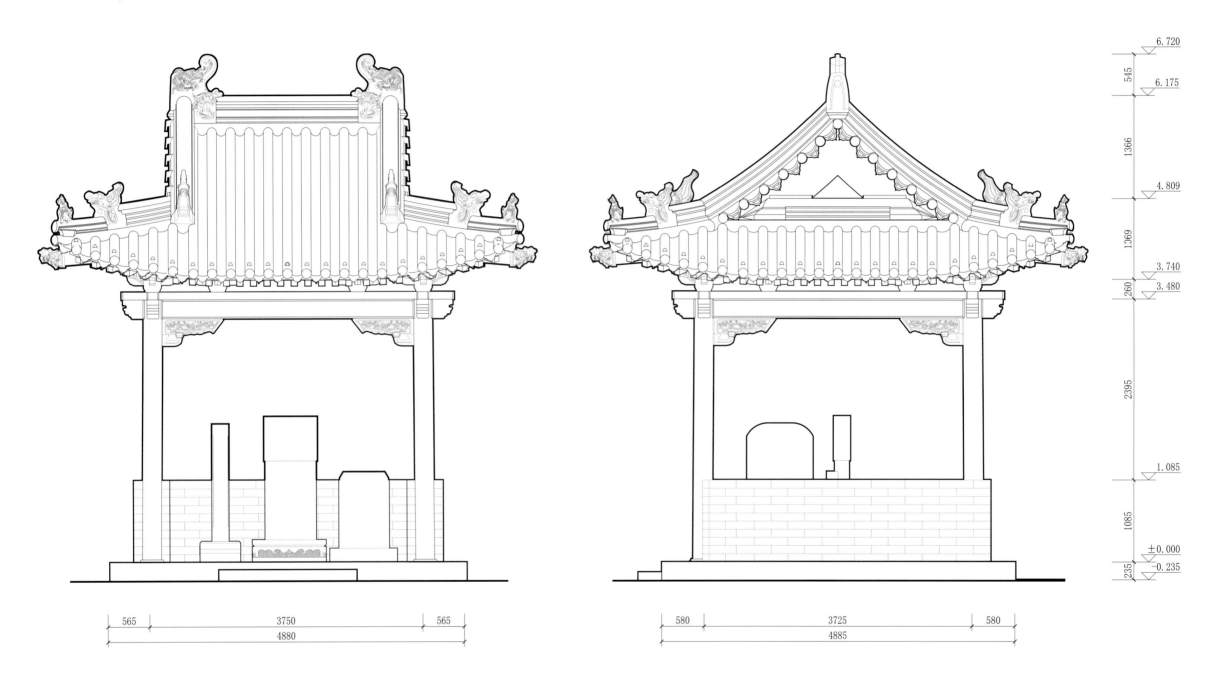

故井碑亭正立面图
Front elevation of stele pavilion of the well at the old residence

故井碑亭侧立面图
Side elevation of stele pavilion of the well at the old residence

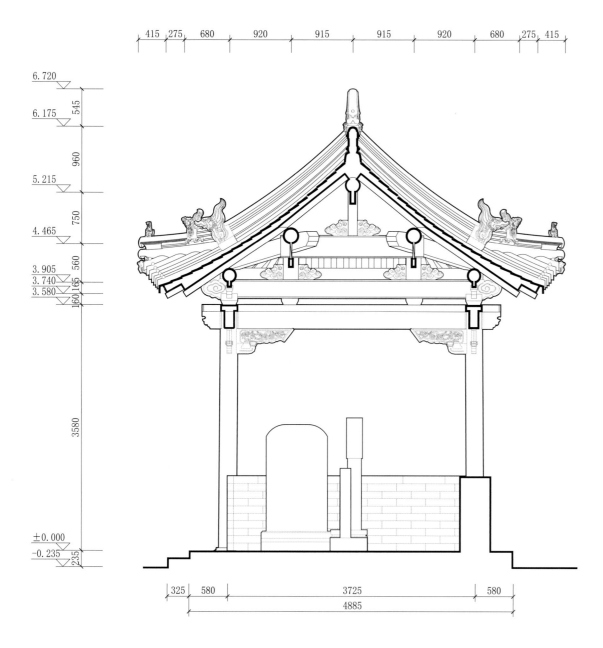

故井碑亭横剖面图
Cross-section of stele pavilion of the well at the old residence

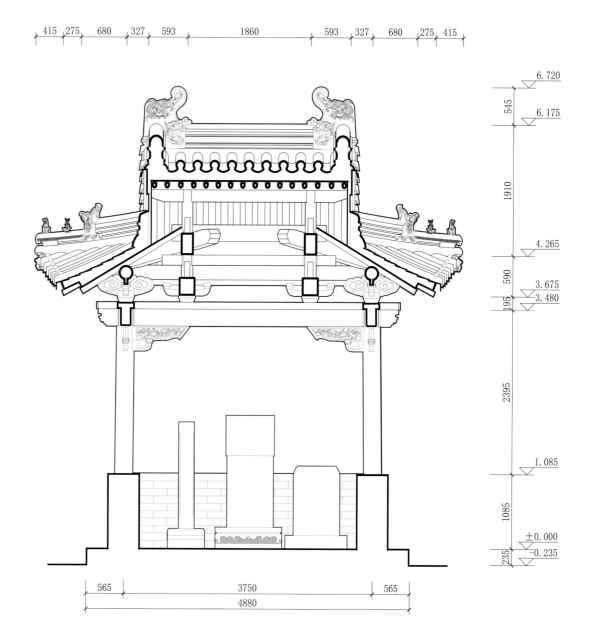

故井碑亭纵剖面图
Longitudinal section of stele pavilion of the well at the old residence

中国古建筑测绘大系·祠庙建筑——曲阜孔庙

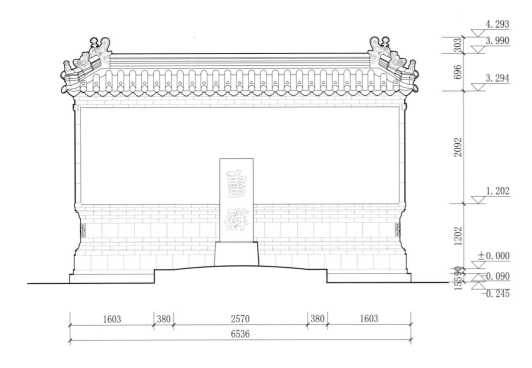

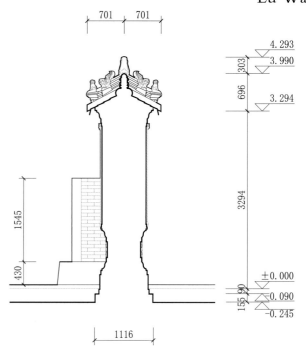

鲁壁正立面图
Front elevation of Lu Wall

鲁壁侧立面图
Side elevation of Lu Wall

鲁壁剖面图
Section of Lu Wall

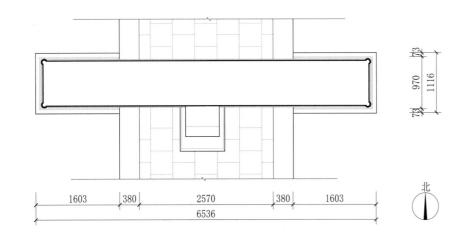

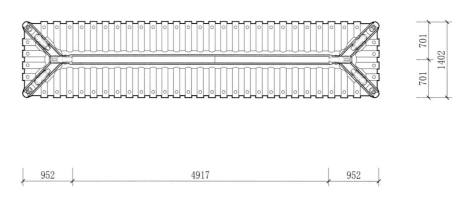

鲁壁平面图
Plan of Lu Wall

鲁壁屋顶平面图
Plan of roof of Lu Wall

崇圣祠门
The Gate of Chongsheng Shrine

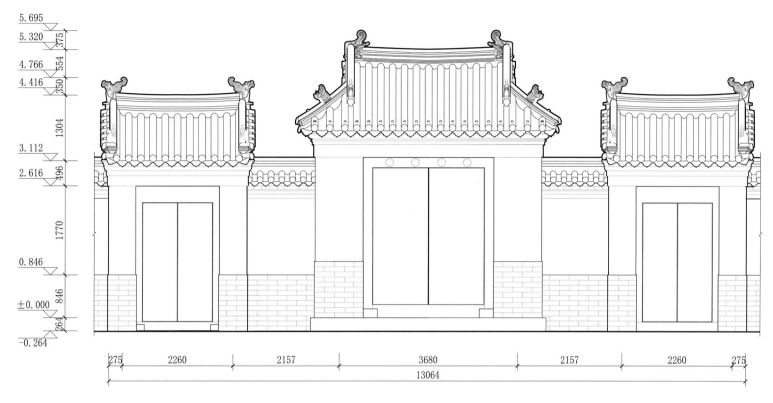

崇圣祠门正立面图
Front elevation of the gate of Chongsheng Shrine

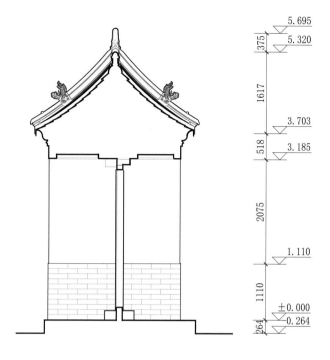

崇圣祠门中门剖面图
Section of the gate of Chongsheng Shrine

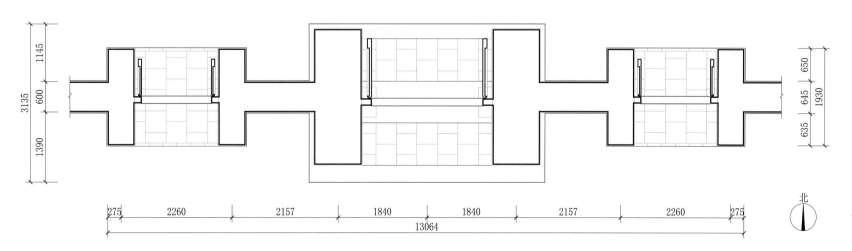

崇圣祠门平面图
Plan of the gate of Chongsheng Shrine

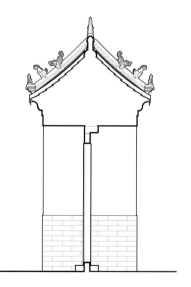

崇圣祠门东门剖面图
Section of the gate of Chongsheng Shrine

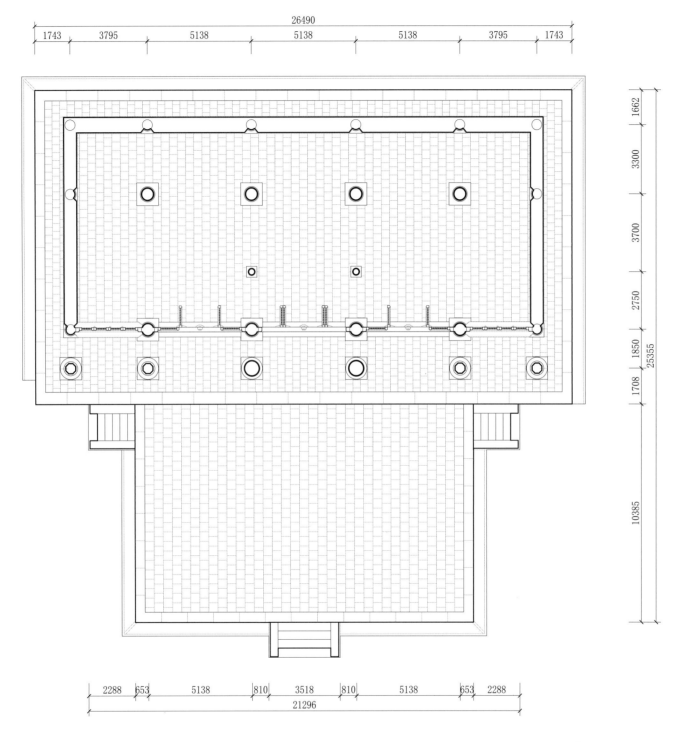

崇圣祠平面图
Plan of Chongsheng Shrine

北

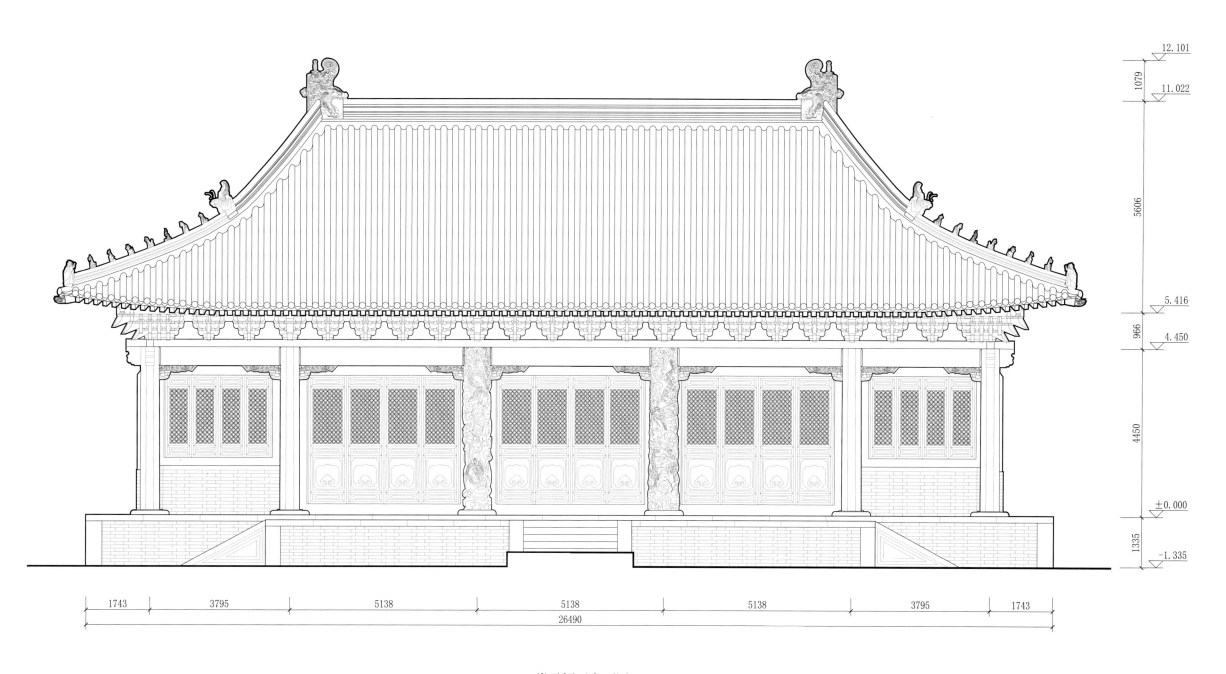

12.101

1079

11.022

5606

5.416

966

4.450

4450

±0.000

1335

−1.335

1743　3795　5138　5138　5138　3795　1743

26490

崇圣祠正立面图
Front elevation of Chongsheng Shrine

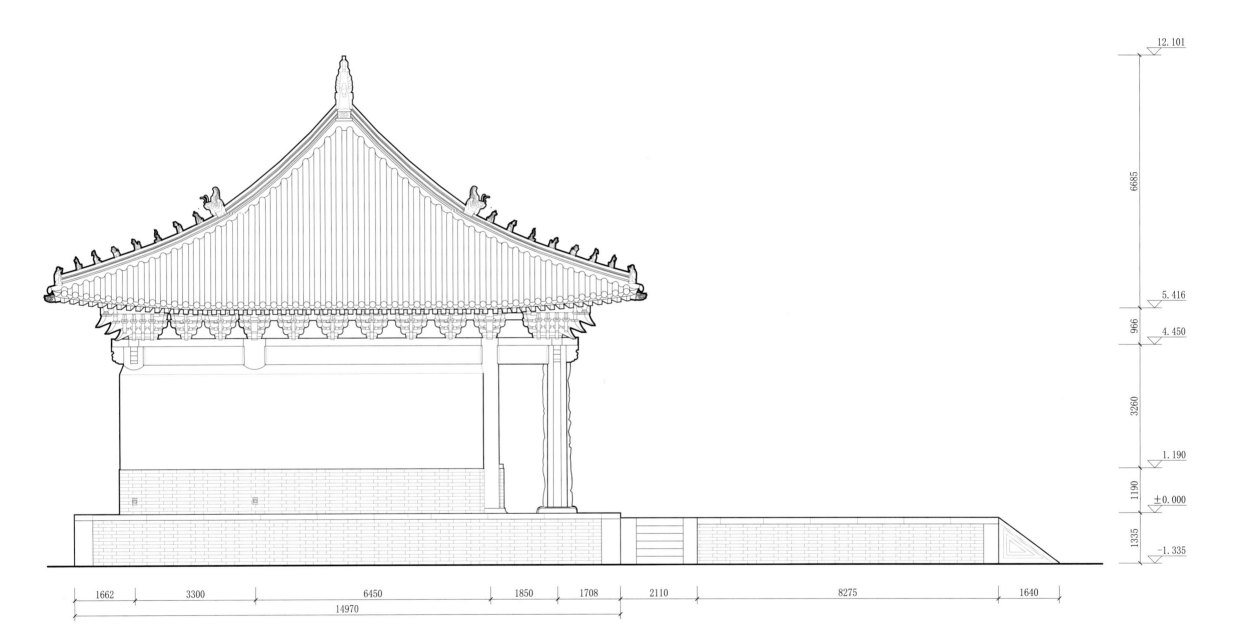

12.101

6685

5.416

196

966

4.450

3260

1.190

1190

±0.000

1335

-1.335

1662　3300　6450　1850　1708　2110　8275　1640

14970

崇圣祠侧立面图
Side elevation of Chongsheng Shrine

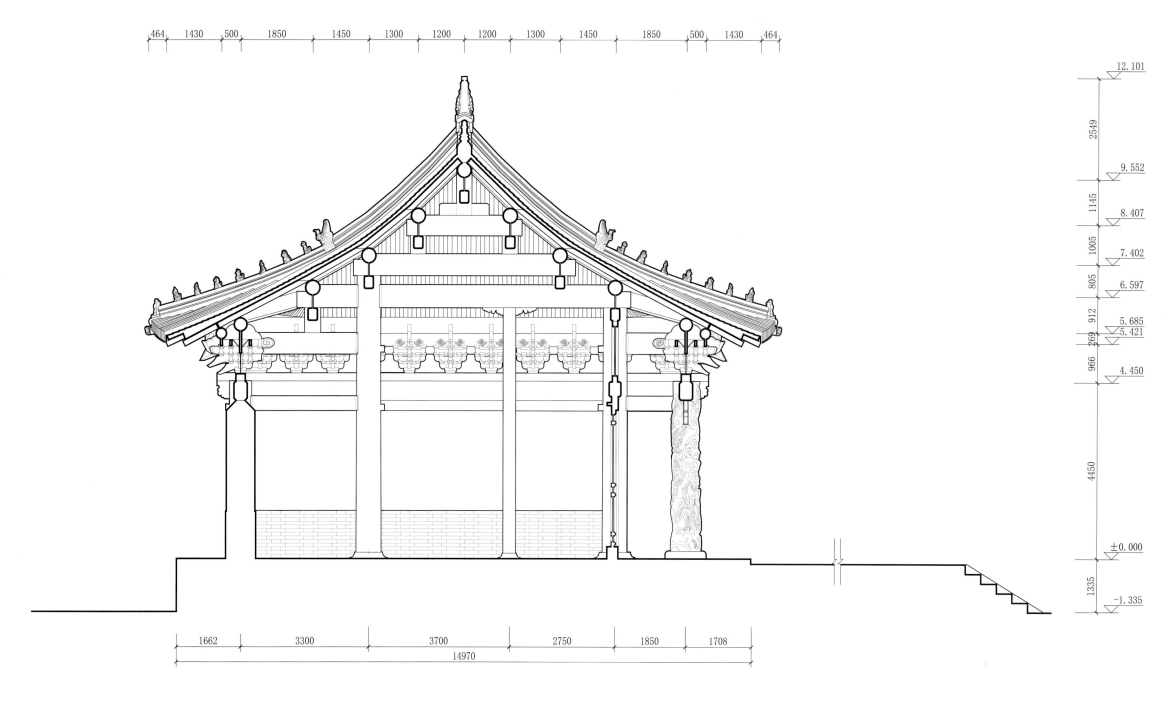

464 1430 500 1850 1450 1300 1200 1200 1300 1450 1850 500 1430 464

12.101
2549
9.552
1145
8.407
1005
7.402
805
6.597
912
5.685
269
5.421
966
4.450
4450
±0.000
1335
-1.335

1662 3300 3700 2750 1850 1708
14970

崇圣祠明间剖面图
Section of central bay of Chongsheng Shrine

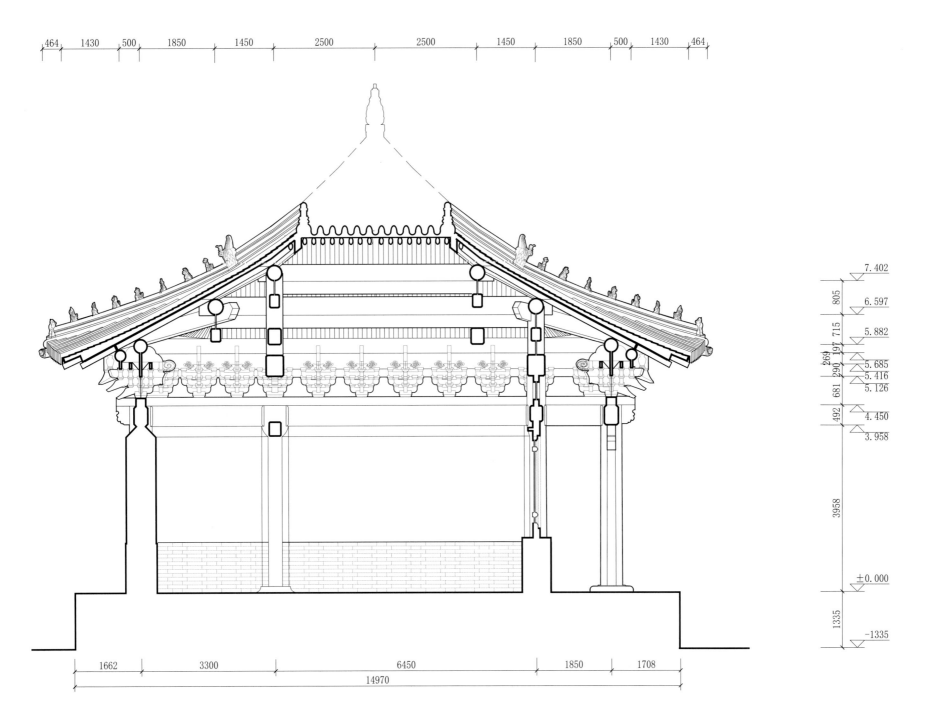

464　1430　500　1850　1450　2500　2500　1450　1850　500　1430　464

7.402

805

6.597

269　197　715

5.882

290

5.685

681

5.416

492

5.126

4.450

3.958

3958

±0.000

1335

-1335

1662　3300　6450　1850　1708

14970

崇圣祠梢间剖面图
Section of second-to-last bay of Chongsheng Shrine

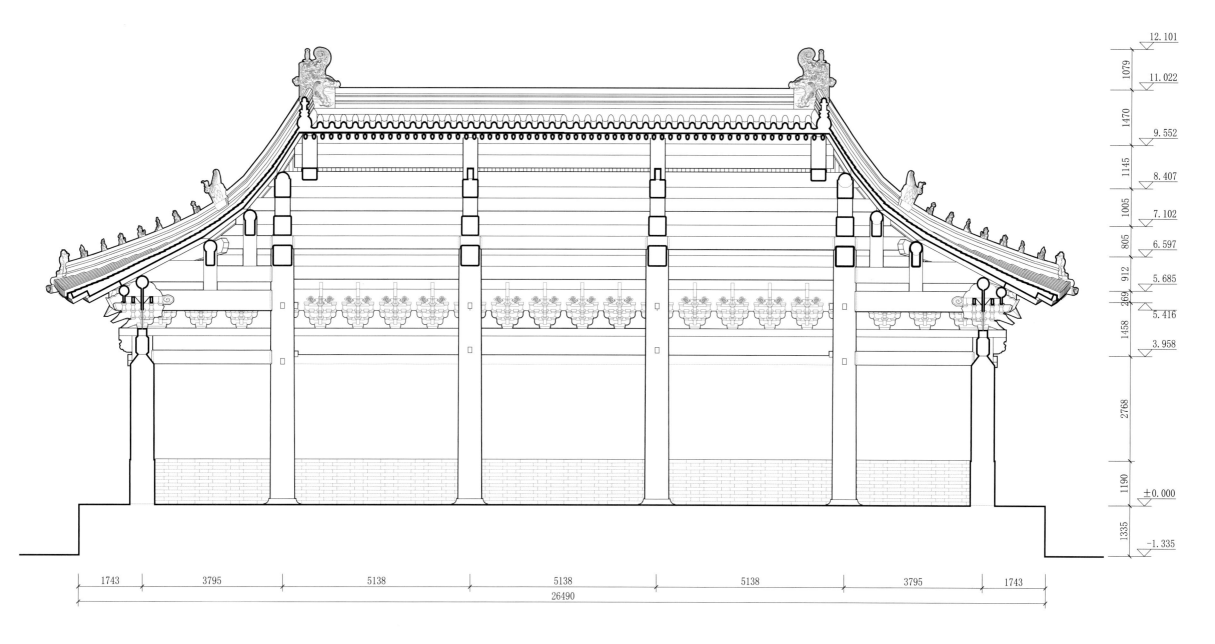

崇圣祠纵剖面图
Longitudinal section of Chongsheng Shrine

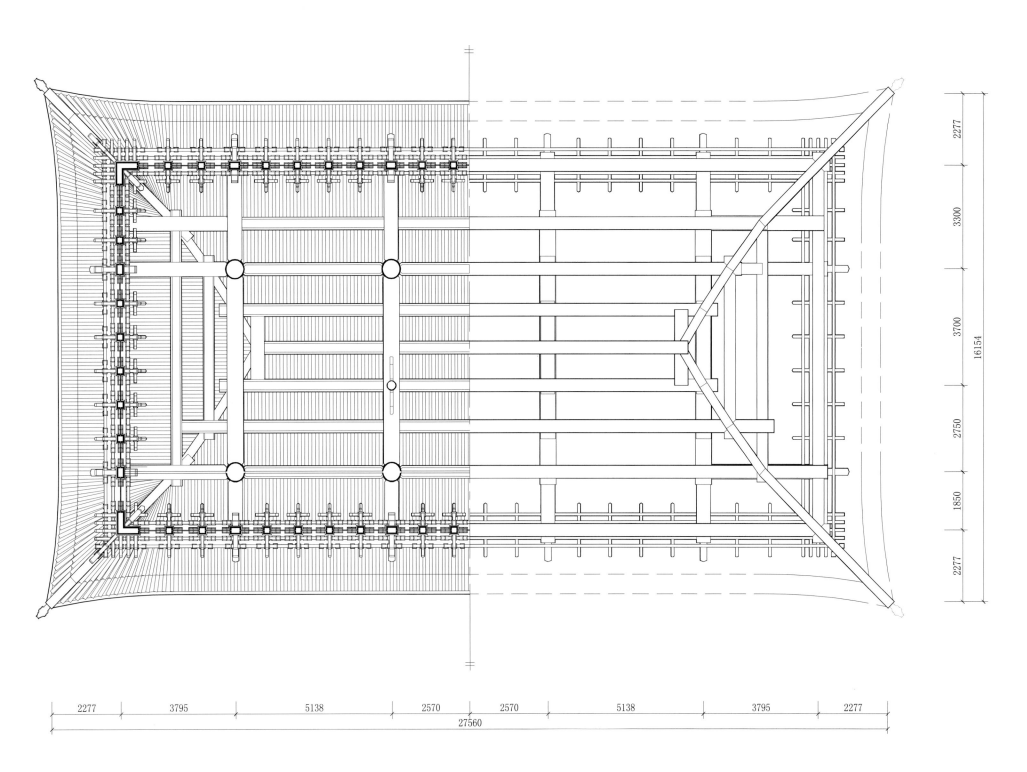

崇圣祠梁架仰视及俯视图

Plan of framework of Chongsheng Shrine as seen from below

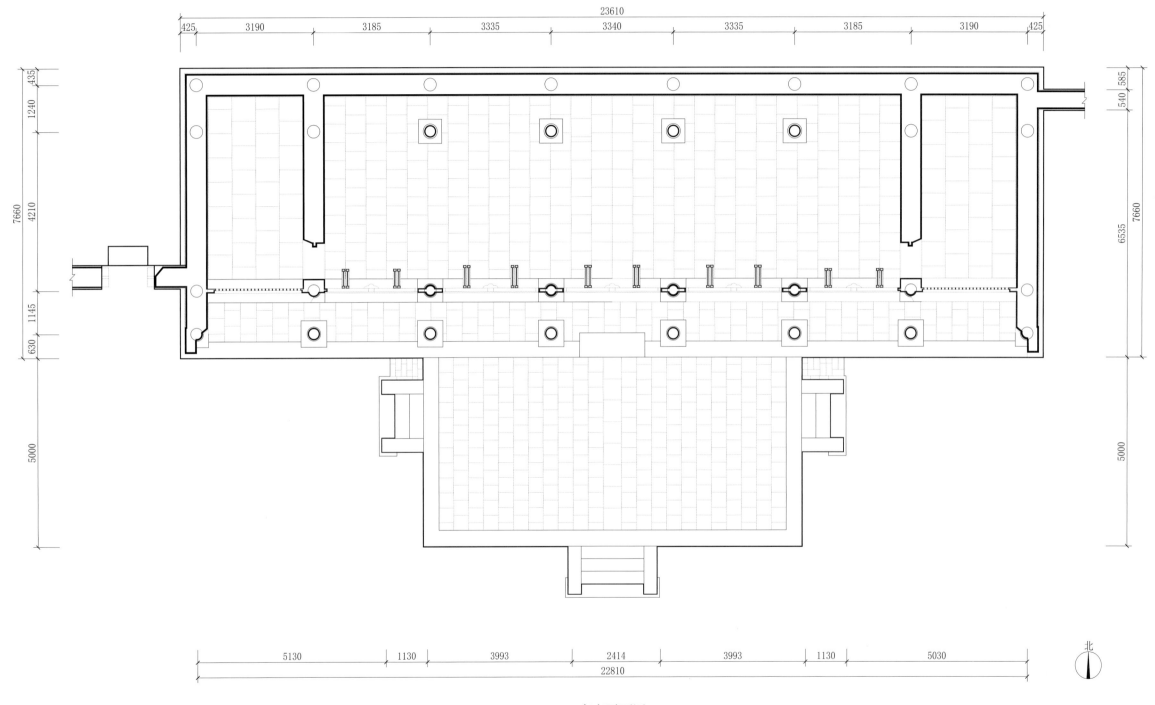

家庙平面图
Plan of the Ancestral Temple

北

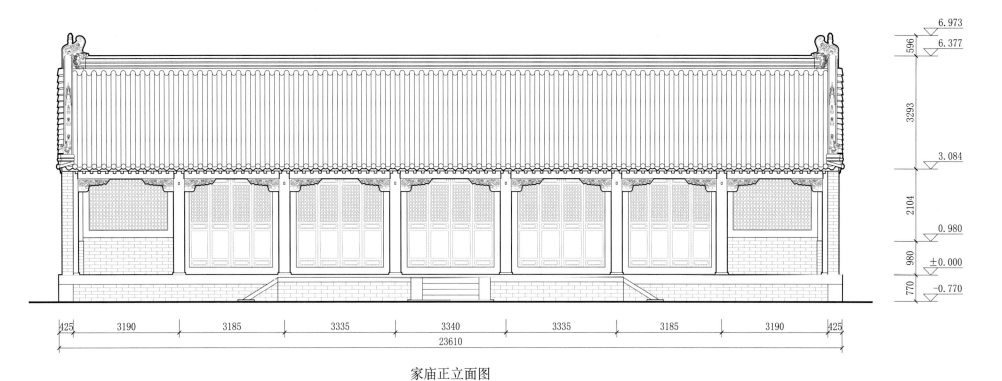

家庙正立面图
Front elevation of the Ancestral Temple

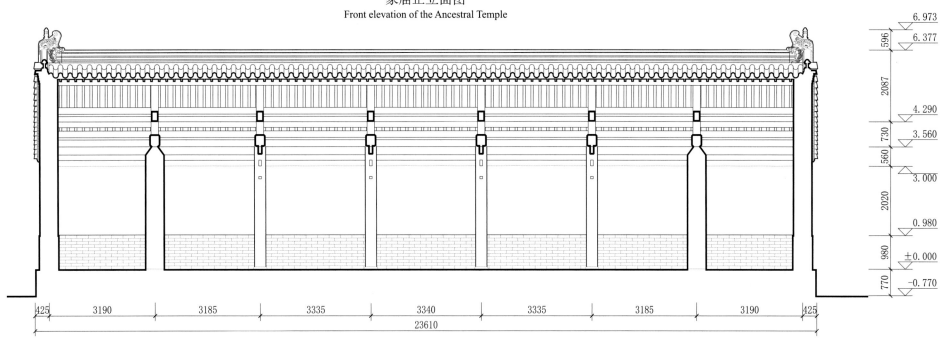

家庙纵剖面图
Longitudinal section of the Ancestral Temple

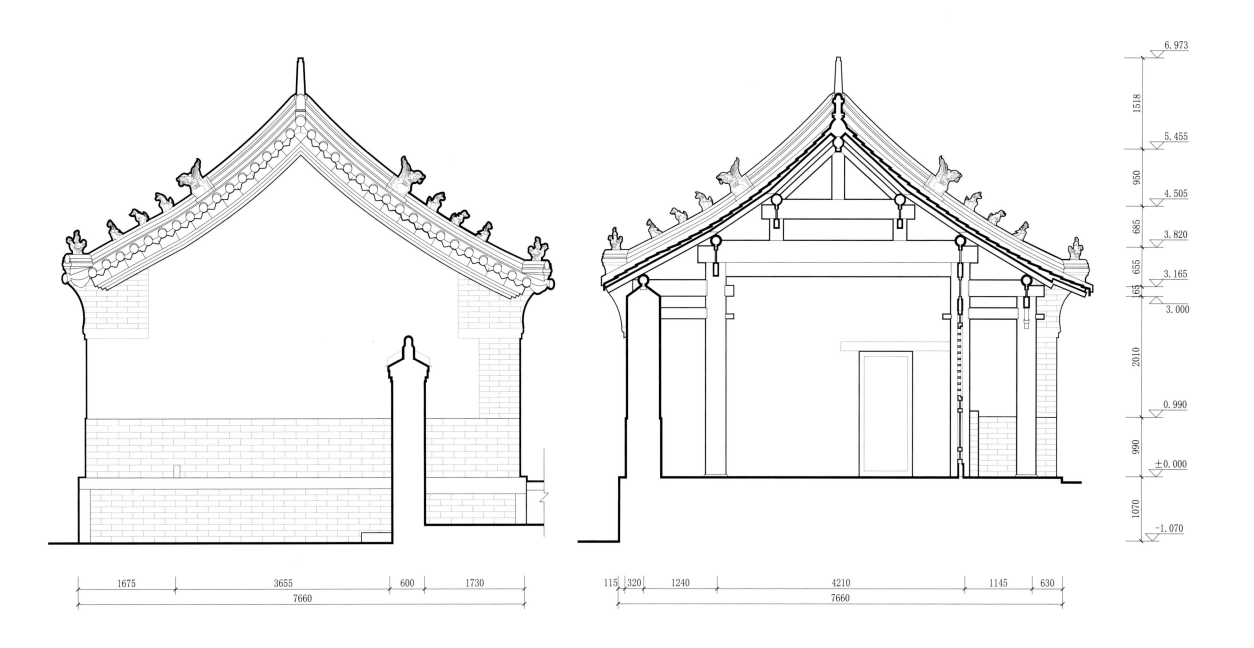

6. 973

1518

5. 455

950

4. 505

685

3. 820

655

3. 165

165

3. 000

2010

0. 990

990

±0. 000

1070

-1. 070

1675　　3655　　600　　1730

7660

115　320　1240　　4210　　1145　630

7660

家庙侧立面图
Side elevation of the Ancestral Temple

家庙横剖面图
Cross-section of the Ancestral Temple

殿庭西院

The West Court

中国古建筑测绘大系·祠庙建筑 —— 曲阜孔庙

206

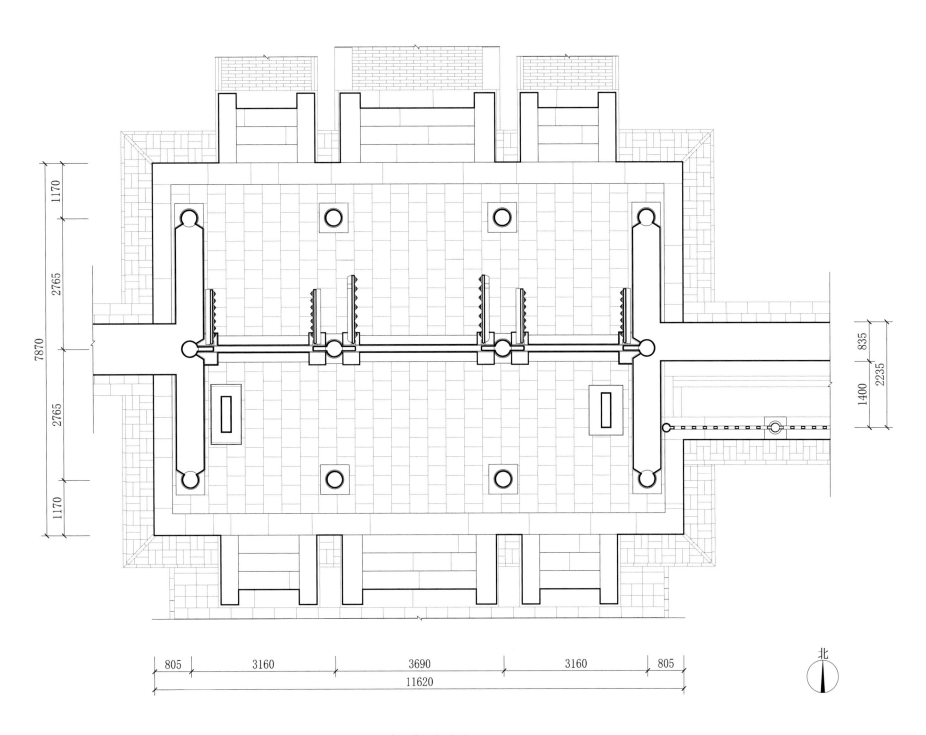

启圣门平面图
Plan of Qisheng Gate

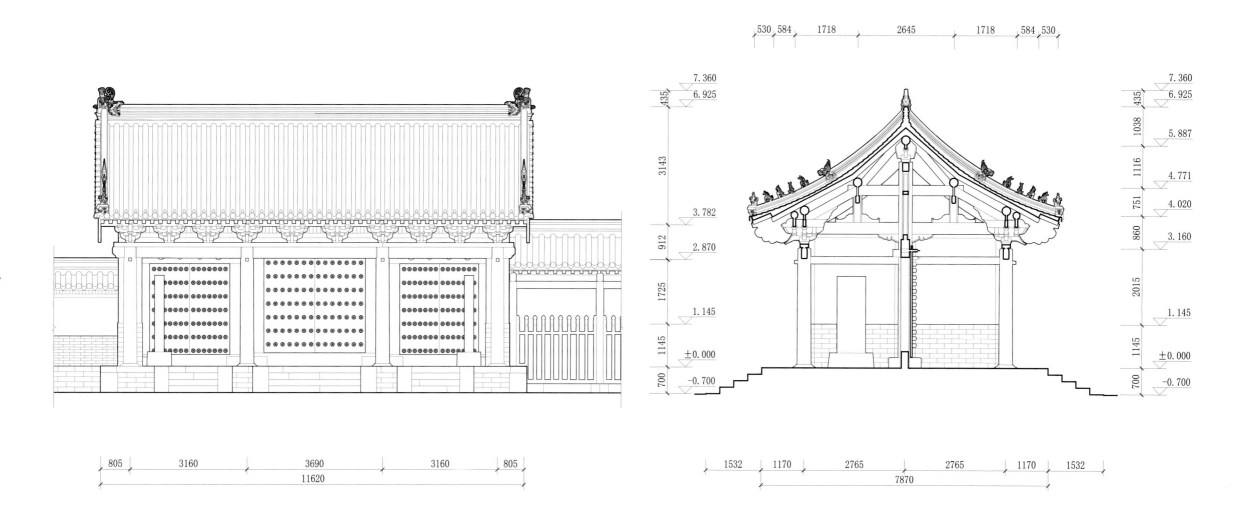

启圣门正立面图
Front elevation of Qisheng Gate

启圣门明间剖面图
Section of central bay of Qisheng Gate

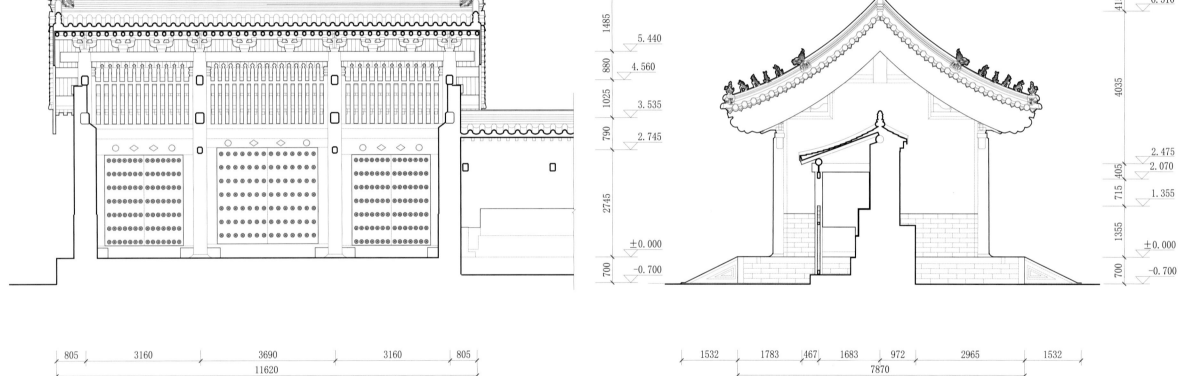

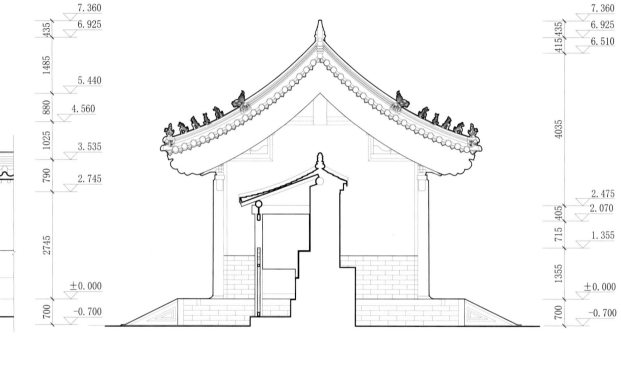

启圣门纵剖面图
Longitudinal section of Qisheng Gate

启圣门侧立面图
Side elevation of Qisheng Gate

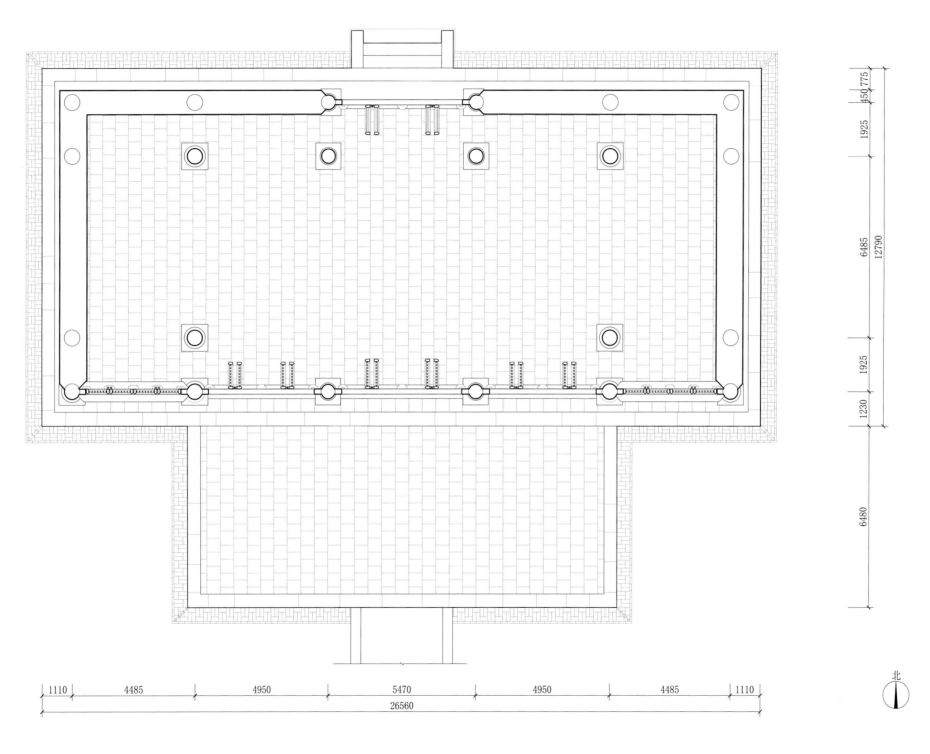

金丝堂平面图
Plan of Jinsi Hall

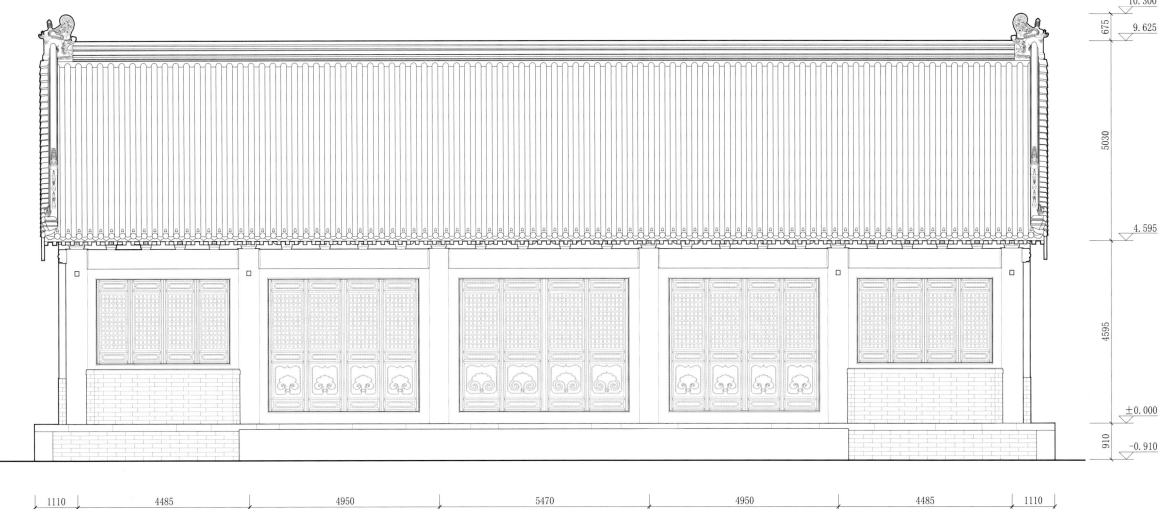

10.300

675

9.625

5030

4.595

210

4595

±0.000

910

−0.910

1110　　4485　　4950　　5470　　4950　　4485　　1110

26560

金丝堂正立面图
Front elevation of Jinsi Hall

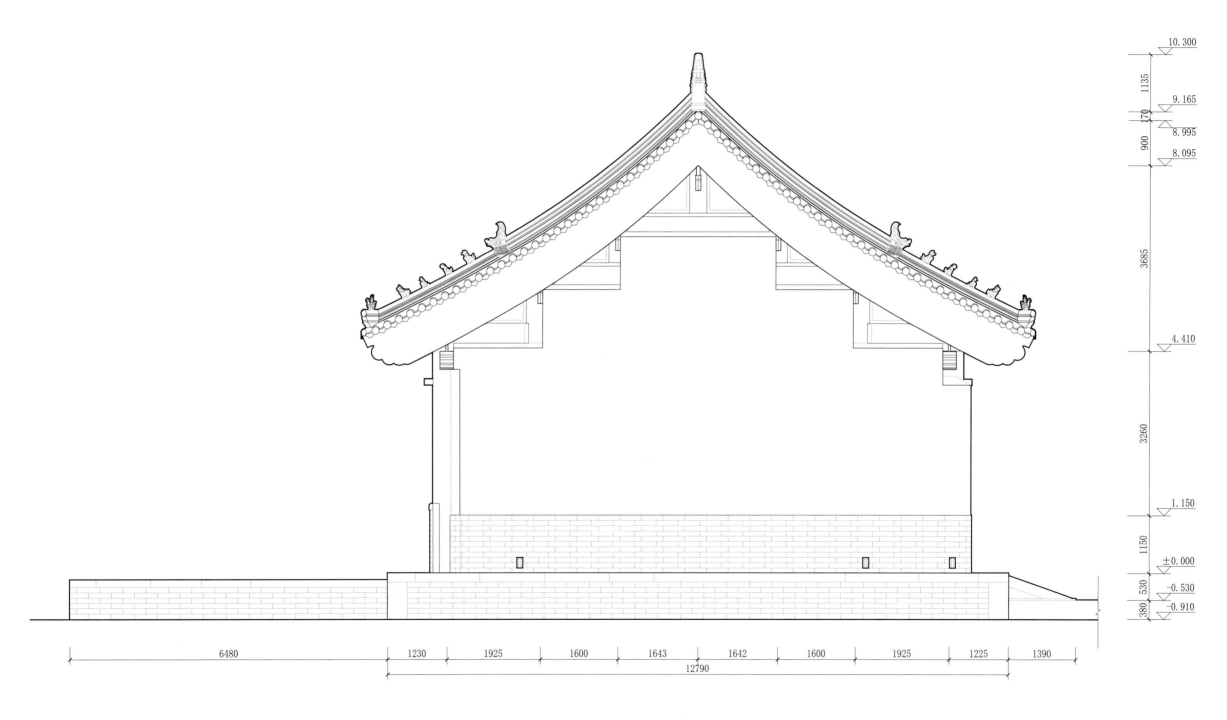

金丝堂侧立面图
Side elevation of Jinsi Hall

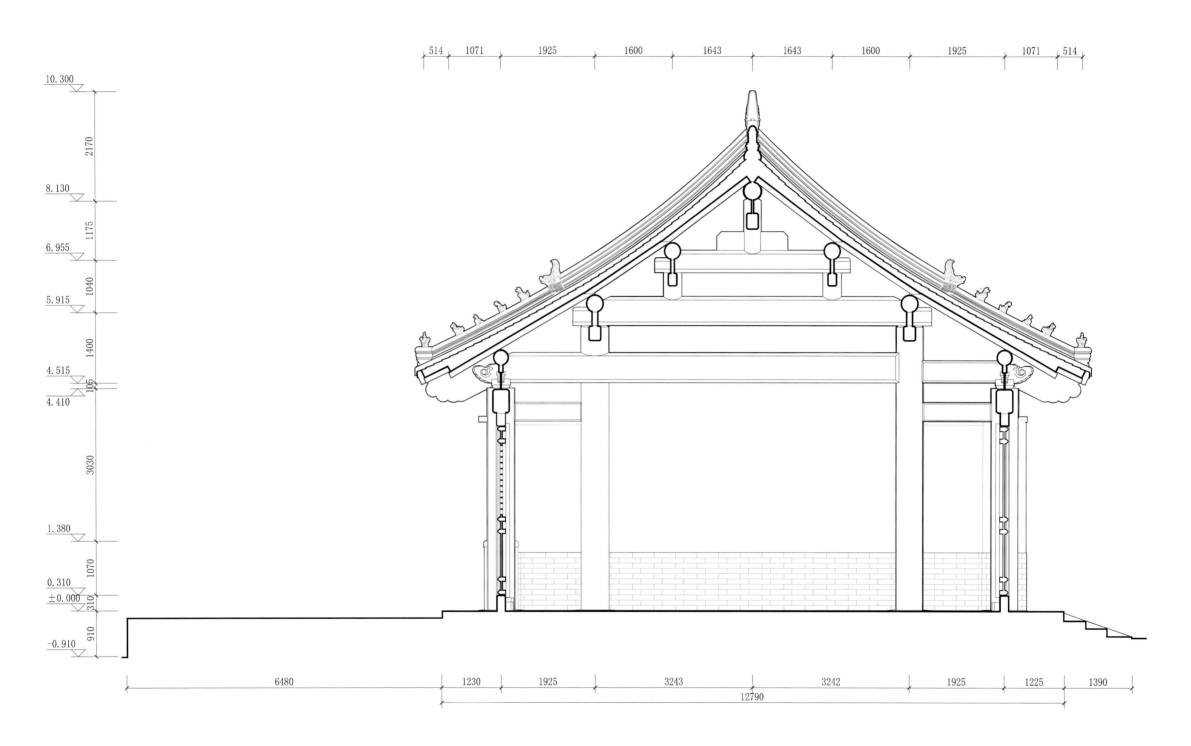

金丝堂明间剖面图
Section of central bay of Jinsi Hall

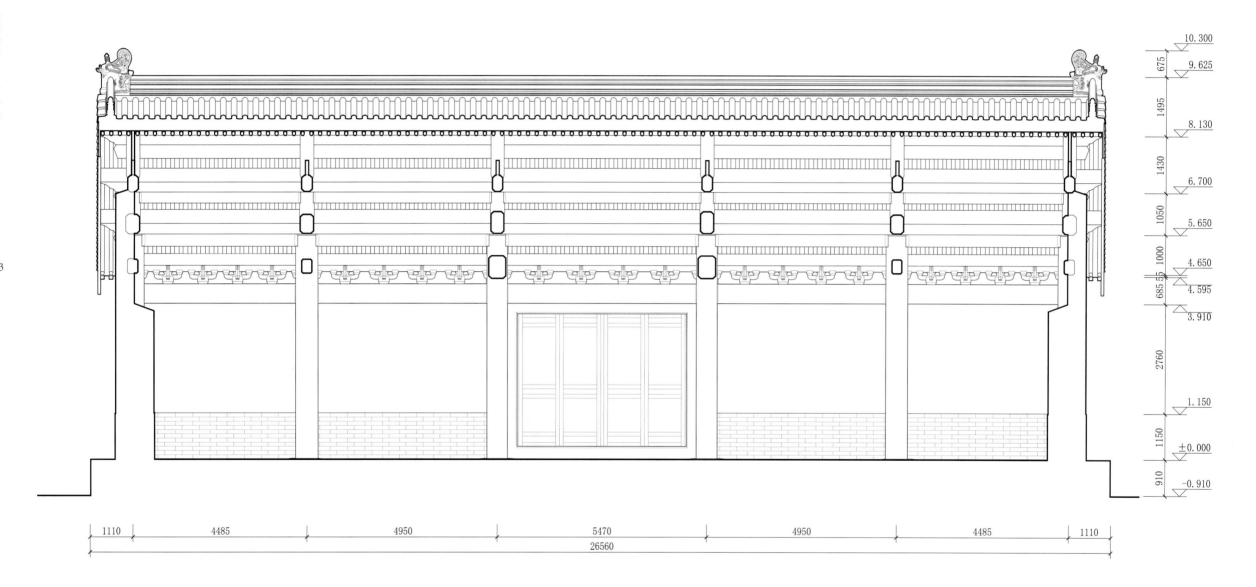

金丝堂纵剖面图
Longitudinal section of Jinsi Hall

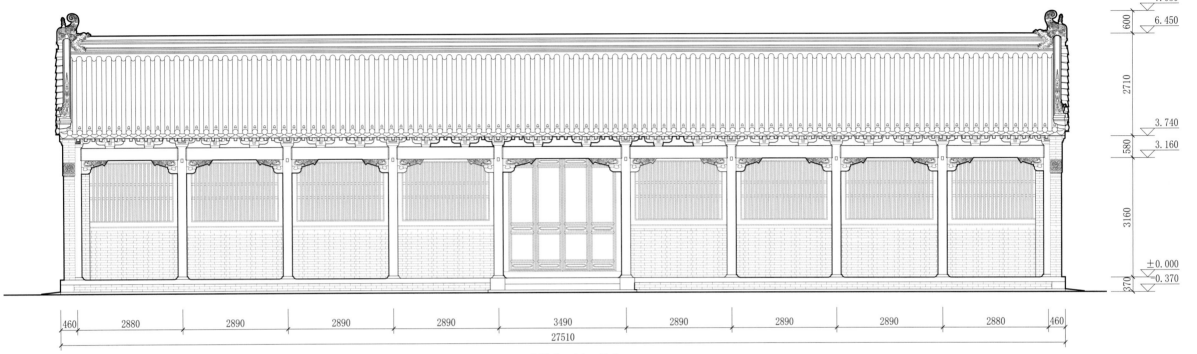

7.050
600
6.450
2710
3.740
580
3.160
3160
±0.000
370
-0.370

460 2880 2890 2890 2890 3490 2890 2890 2890 2880 460
27510

乐器库正立面图
Front elevation of musical instruments storeroom

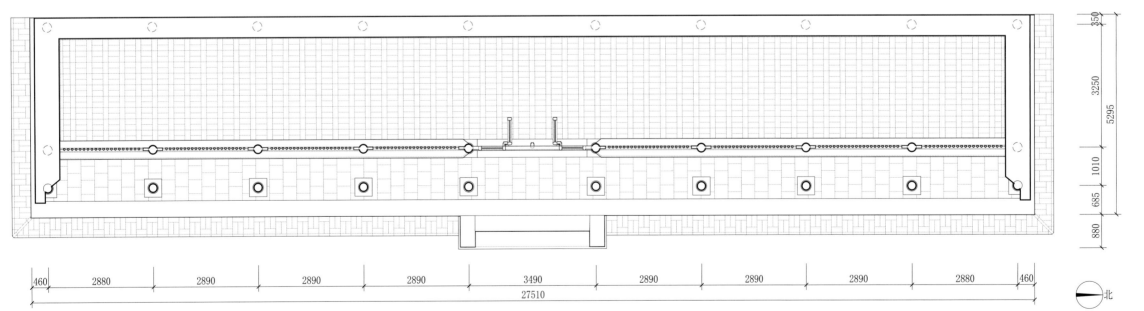

350
3250
5295
1010
685
880

460 2880 2890 2890 2890 3490 2890 2890 2890 2880 460
27510

北

乐器库平面图
Plan of musical instruments storeroom

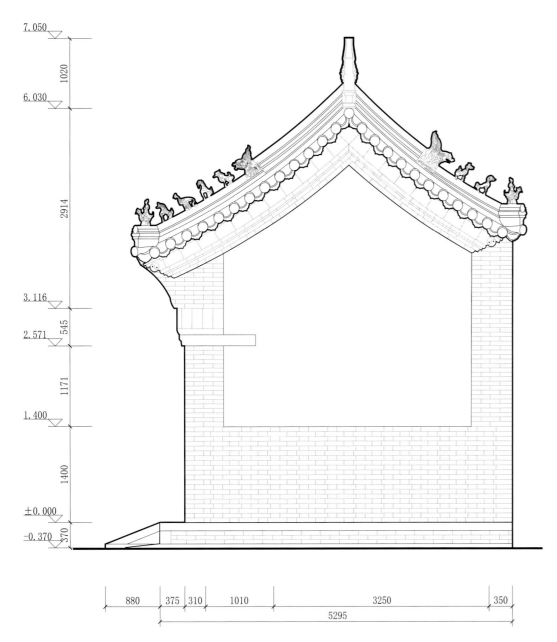

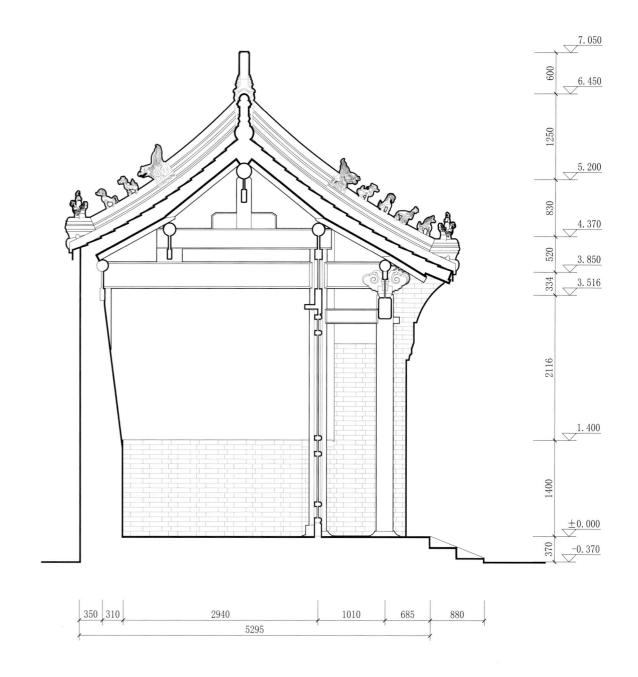

乐器库侧立面图
Side elevation of musical instruments storeroom

乐器库横剖面图
Section of musical instruments storeroom

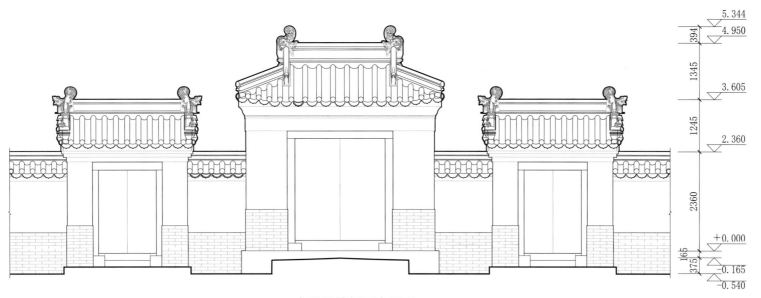

启圣殿院门正立面图
Front elevation of the gate of courtyard with Qisheng Hall

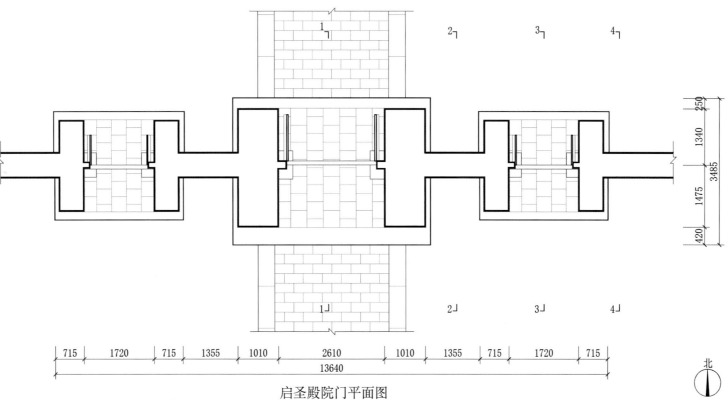

启圣殿院门平面图
Plan of the gate of courtyard with Qisheng Hall

北

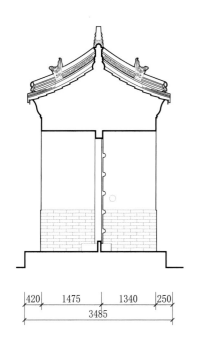

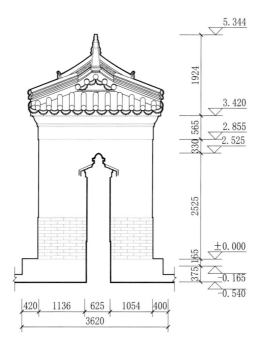

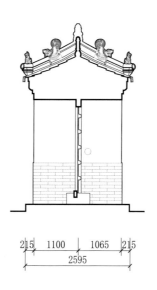

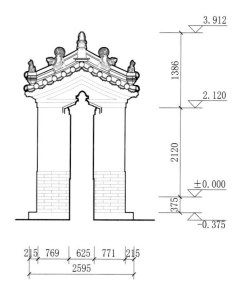

启圣殿院门 1—1 剖面图
Section 1-1 of the gate of courtyard with Qisheng Hall

启圣殿院门 2—2 剖面图
Section 2-2 of the gate of courtyard with Qisheng Hall

启圣殿院门 3—3 剖面图
Section 3-3 of the gate of courtyard with Qisheng Hall

启圣殿院门 4—4 剖面图
Section 4-4 of the gate of courtyard with Qisheng Hall

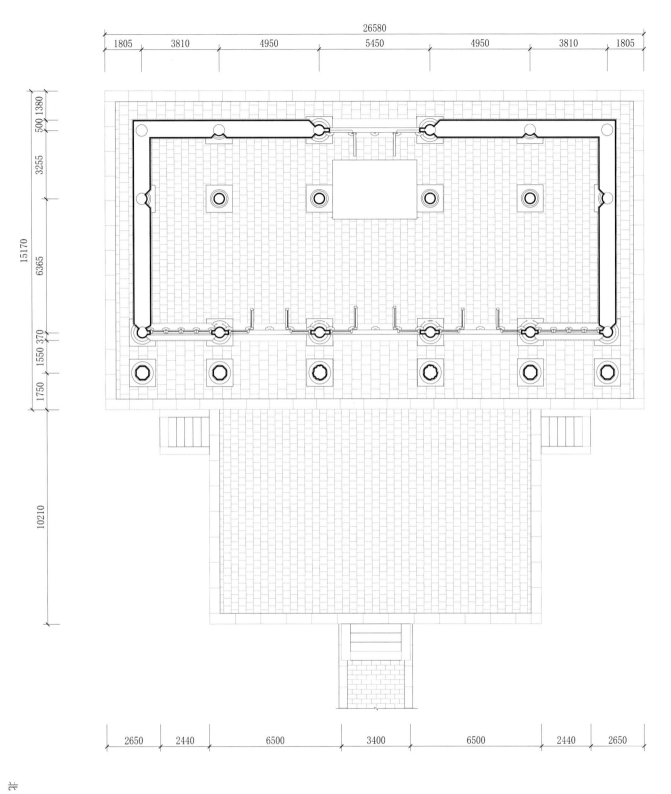

启圣殿平面图
Plan of Qisheng Hall

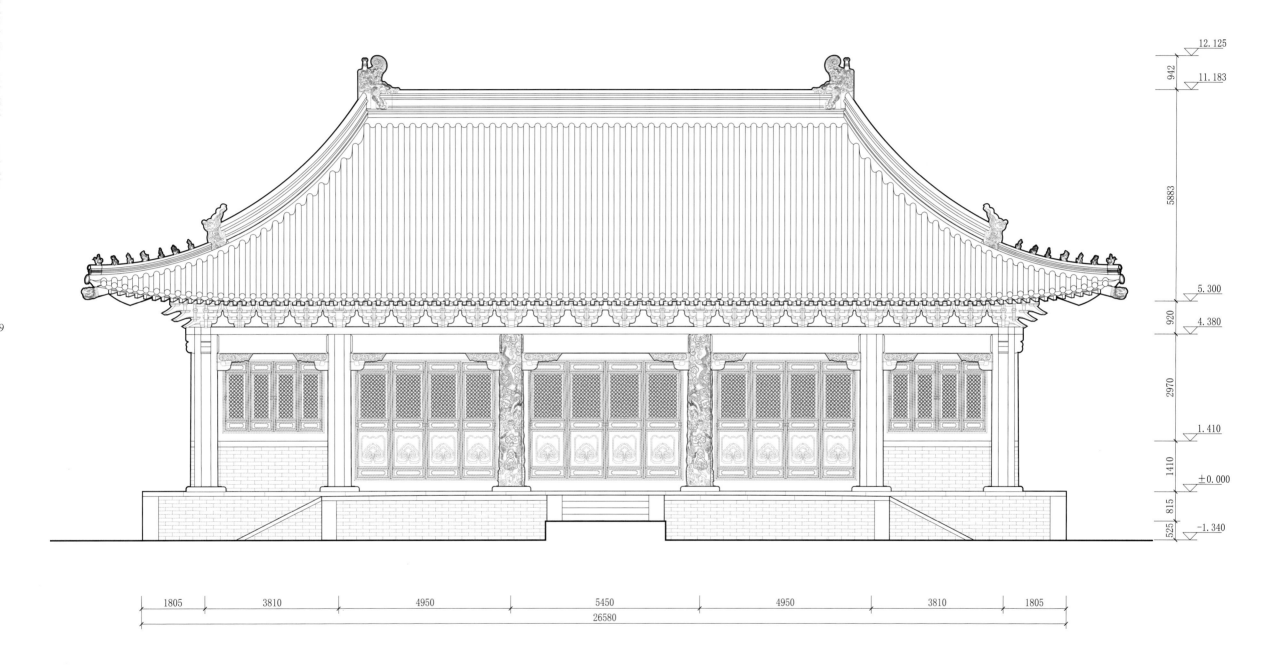

启圣殿正立面图
Front elevation of Qisheng Hall

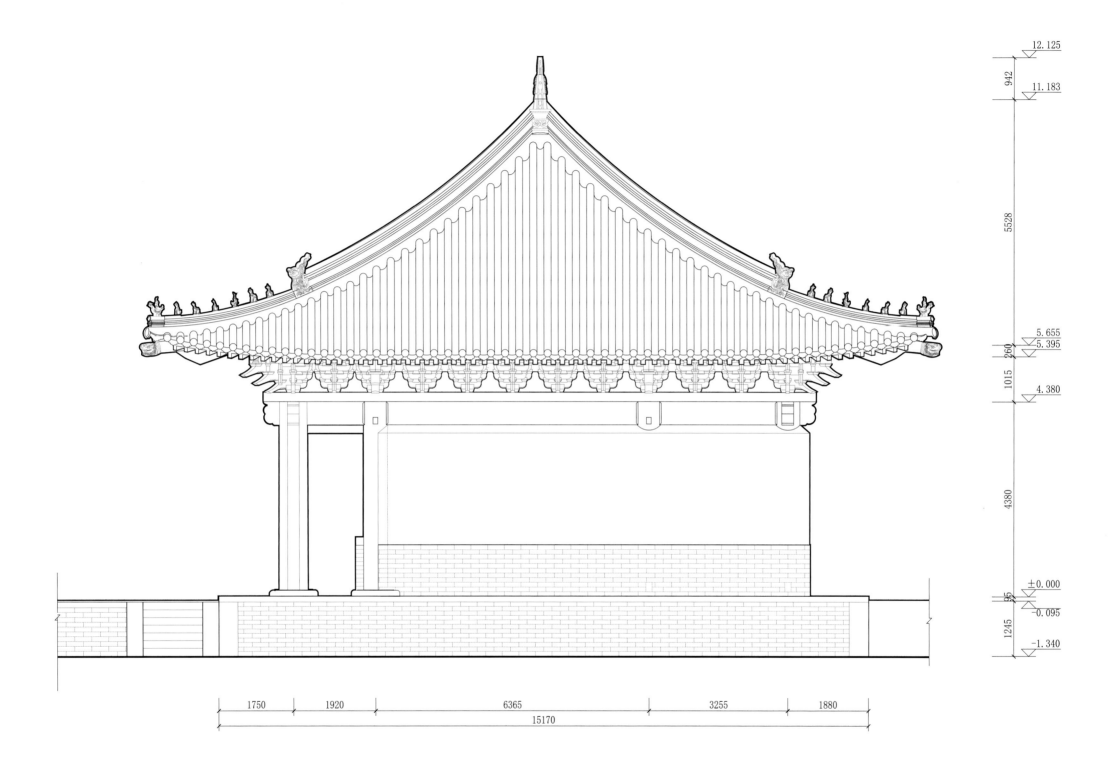

12.125

942

11.183

5528

5.655
5.395

260

1015

4.380

4380

±0.000

−0.095

1245

−1.340

1750　1920　6365　3255　1880

15170

启圣殿侧立面图
Side elevation of Qisheng Hall

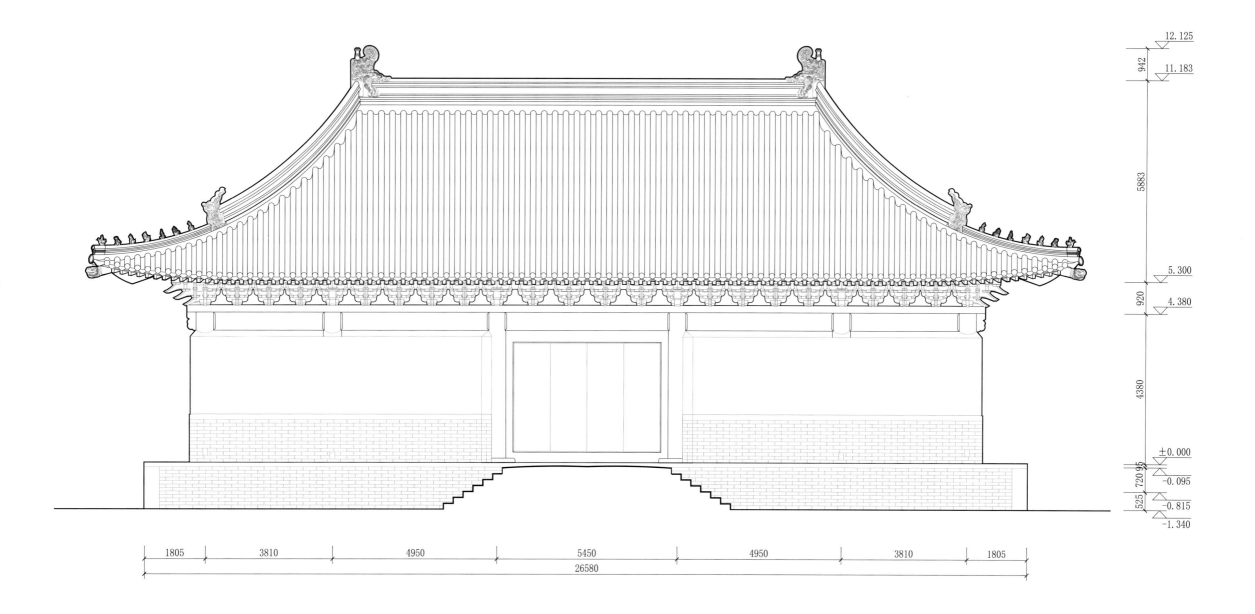

12.125

942

11.183

5883

5.300

920

4.380

4380

±0.000

720 95

−0.095

525

−0.815

−1.340

| 1805 | 3810 | 4950 | 5450 | 4950 | 3810 | 1805 |

26580

启圣殿背立面图
Rear elevation of Qisheng Hall

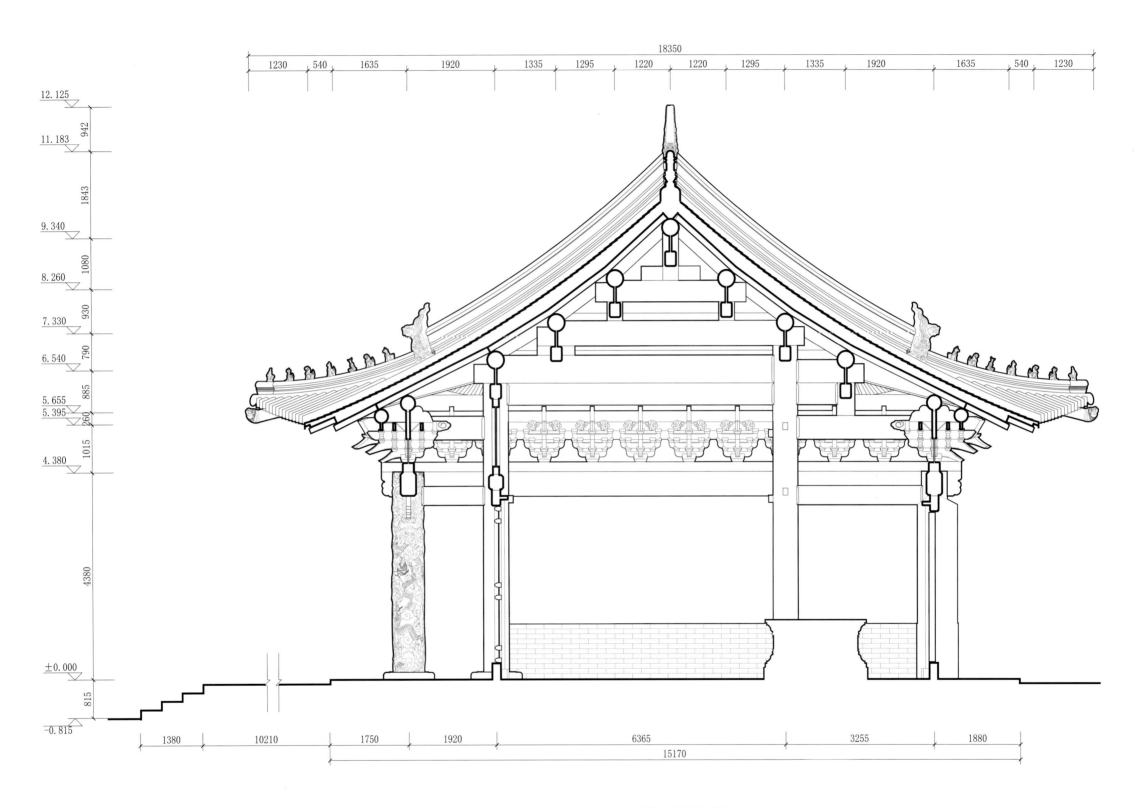

启圣殿明间剖面图
Section of central bay of Qisheng Hall

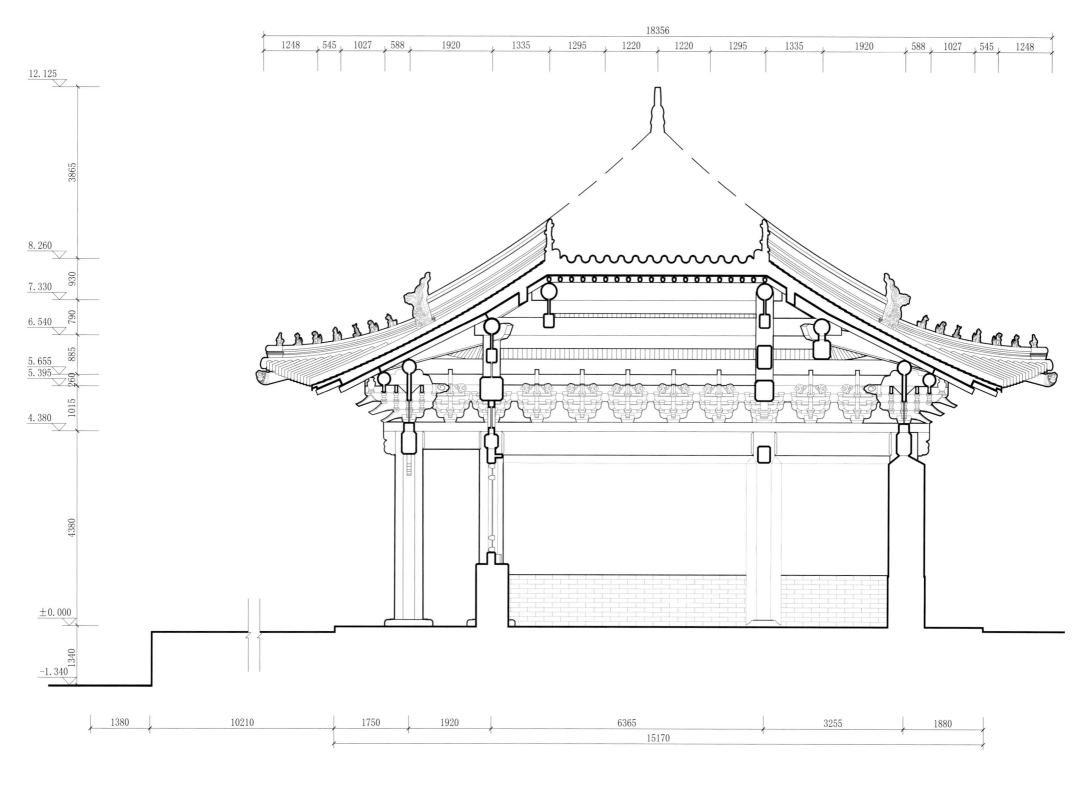

启圣殿梢间剖面图
Section of second-to-last bay of Qisheng Hall

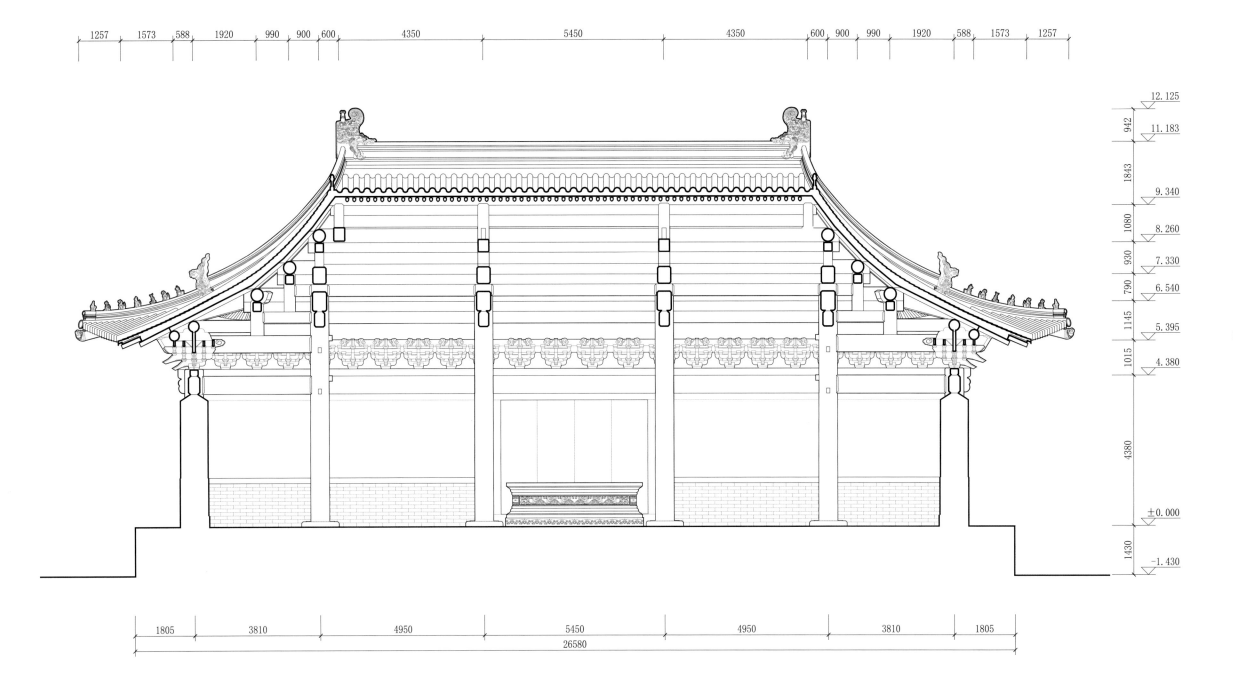

启圣殿纵剖面图
Longitudinal section of Qisheng Hall

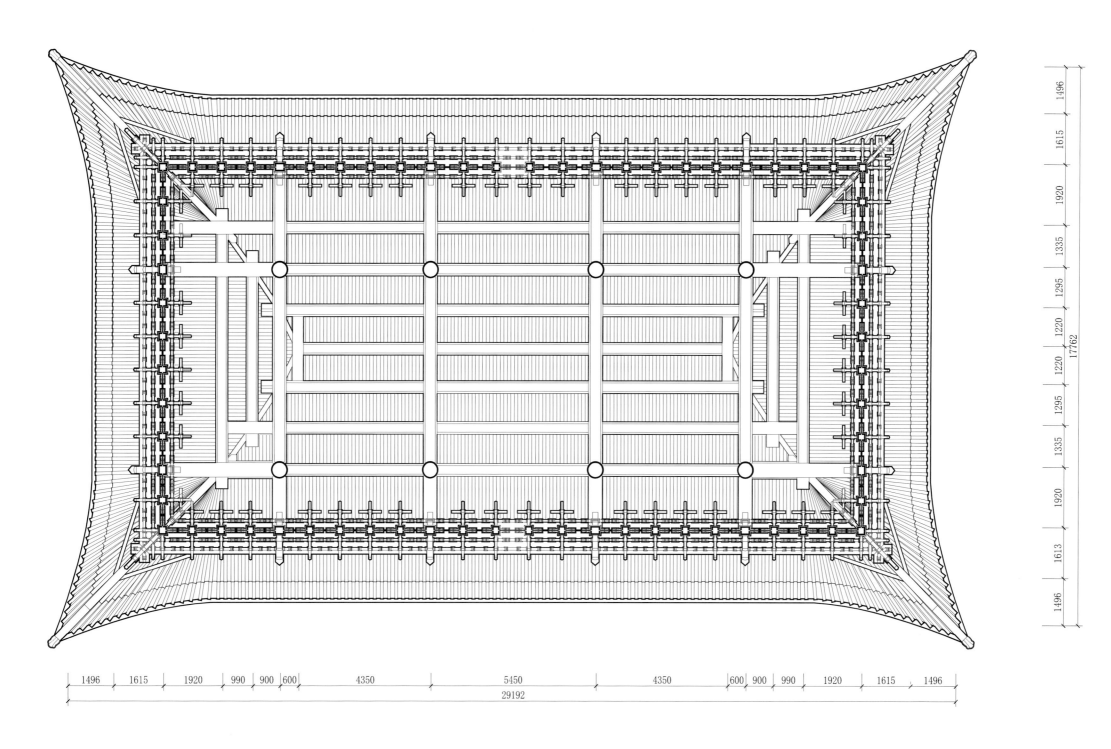

启圣殿梁架仰视图
Plan of framework of Qisheng Hall as seen from below

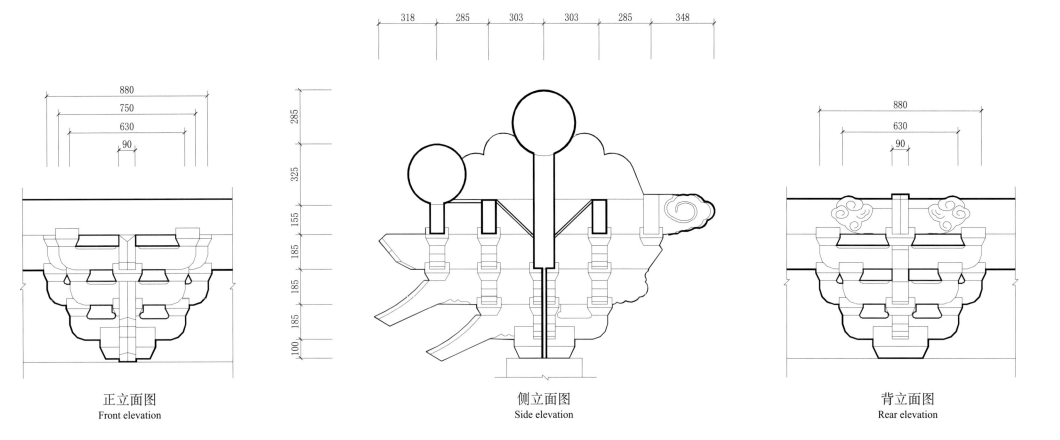

880
750
630
90

正立面图
Front elevation

318　285　303　303　285　348

285
325
155
185
185
185
100

侧立面图
Side elevation

880
630
90

背立面图
Rear elevation

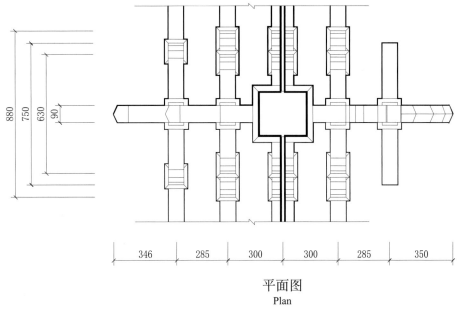

880
750
630
90

346　285　300　300　285　350

平面图
Plan

北

斗栱测量位置

启圣殿平身科斗栱大样图
Intercolumnar bracket sets of Qisheng Hall

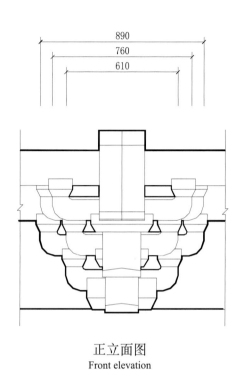

正立面图
Front elevation

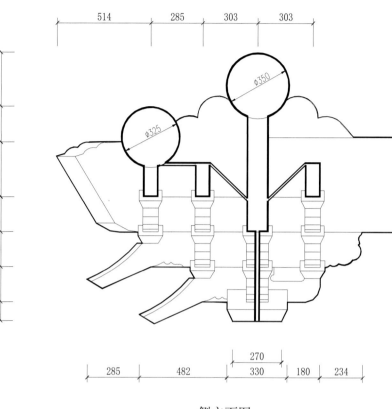

侧立面图
Side elevation

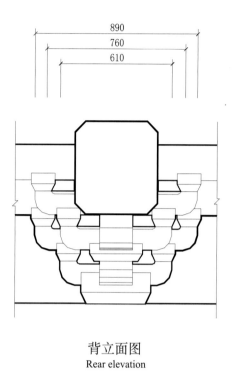

背立面图
Rear elevation

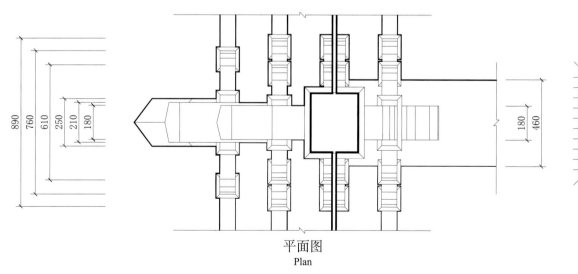

平面图
Plan

启圣殿柱头科斗栱大样图
Bracket sets atop columns of Qisheng Hall

斗栱测量位置

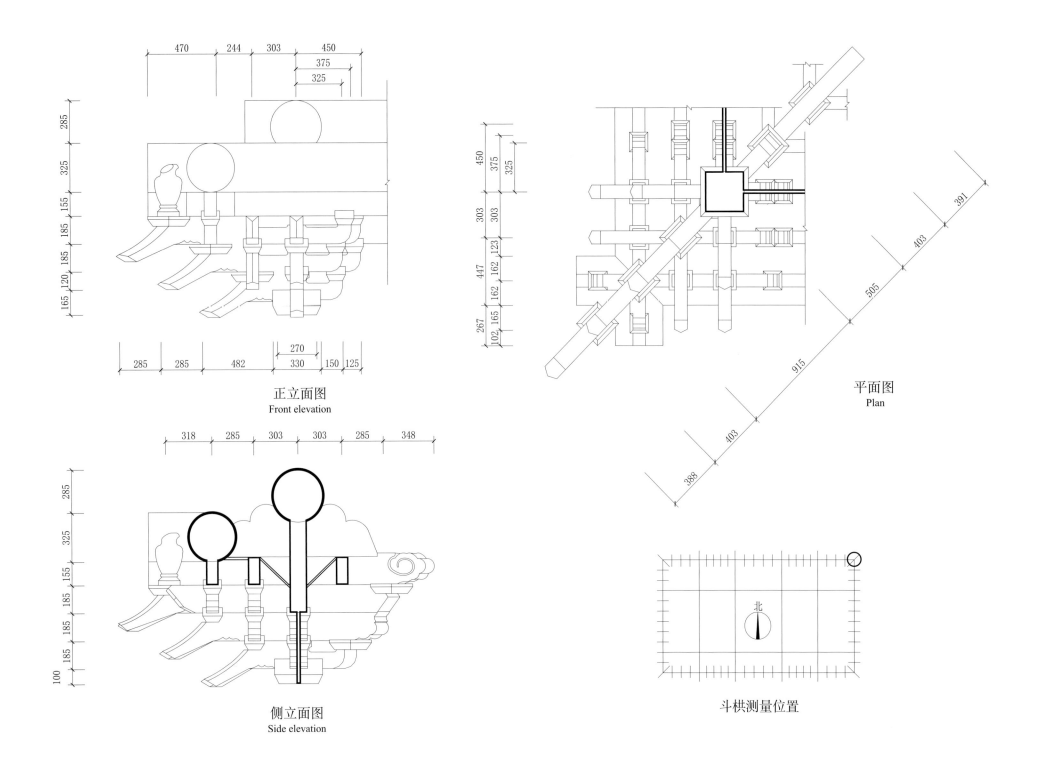

正立面图
Front elevation

平面图
Plan

侧立面图
Side elevation

北

斗栱测量位置

启圣殿角科斗栱大样图
Corner bracket sets of Qisheng Hall

启圣寝殿
Qisheng Resting Hall

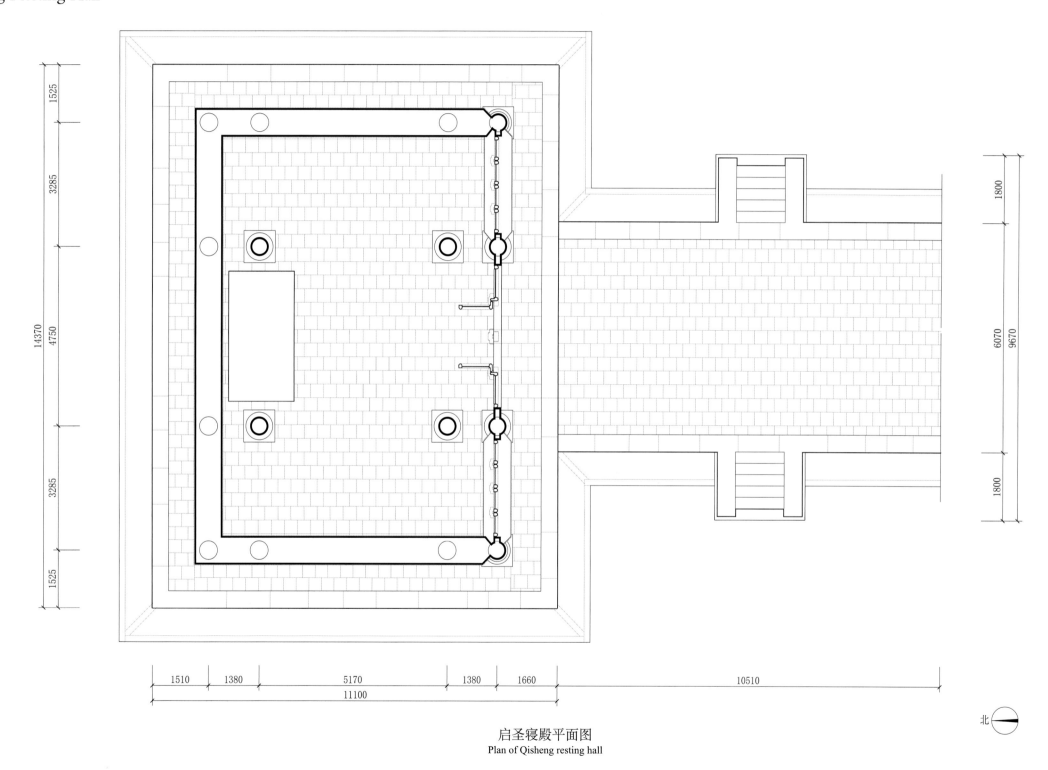

启圣寝殿平面图
Plan of Qisheng resting hall

北

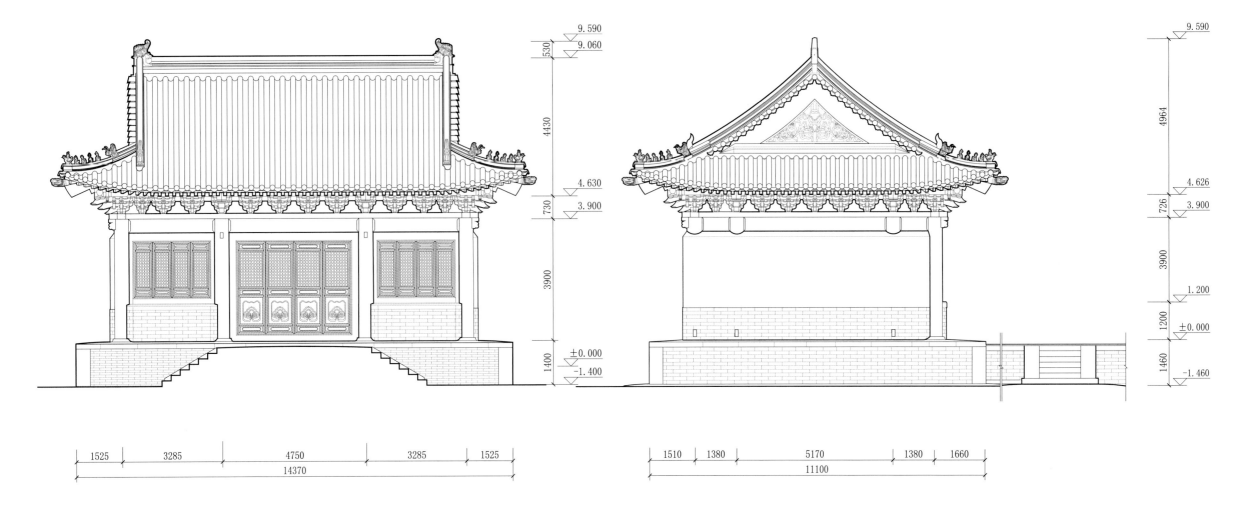

启圣寝殿正立面图
Front elevation of Qisheng resting hall

启圣寝殿侧立面图
Side elevation of Qisheng resting hall

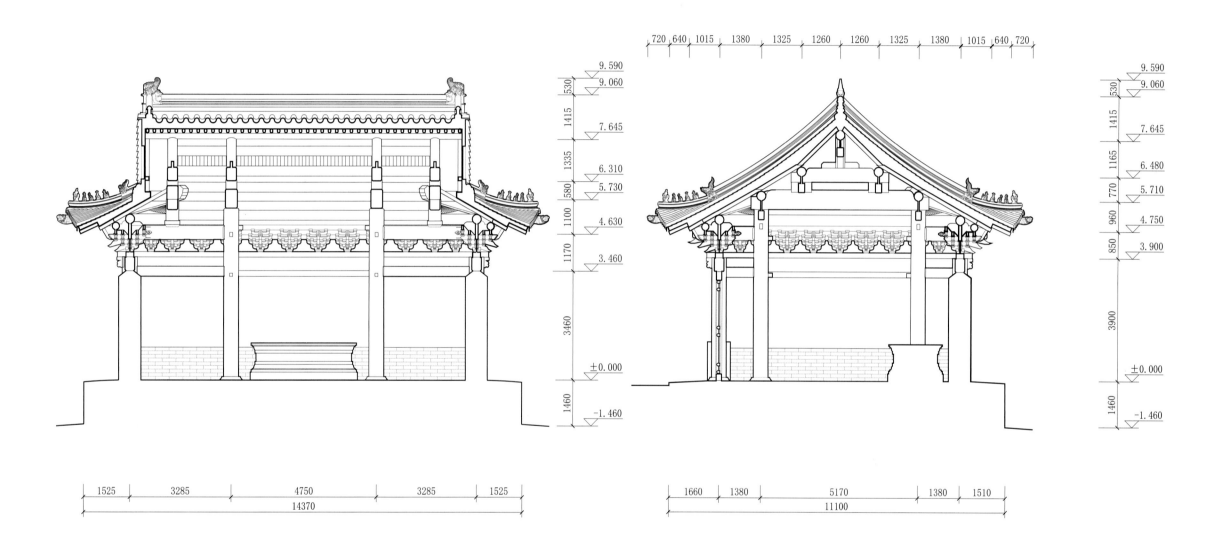

启圣寝殿纵剖面图
Longitudinal section of Qisheng resting hall

启圣寝殿明间剖面图
Section of central bay of Qisheng resting hall

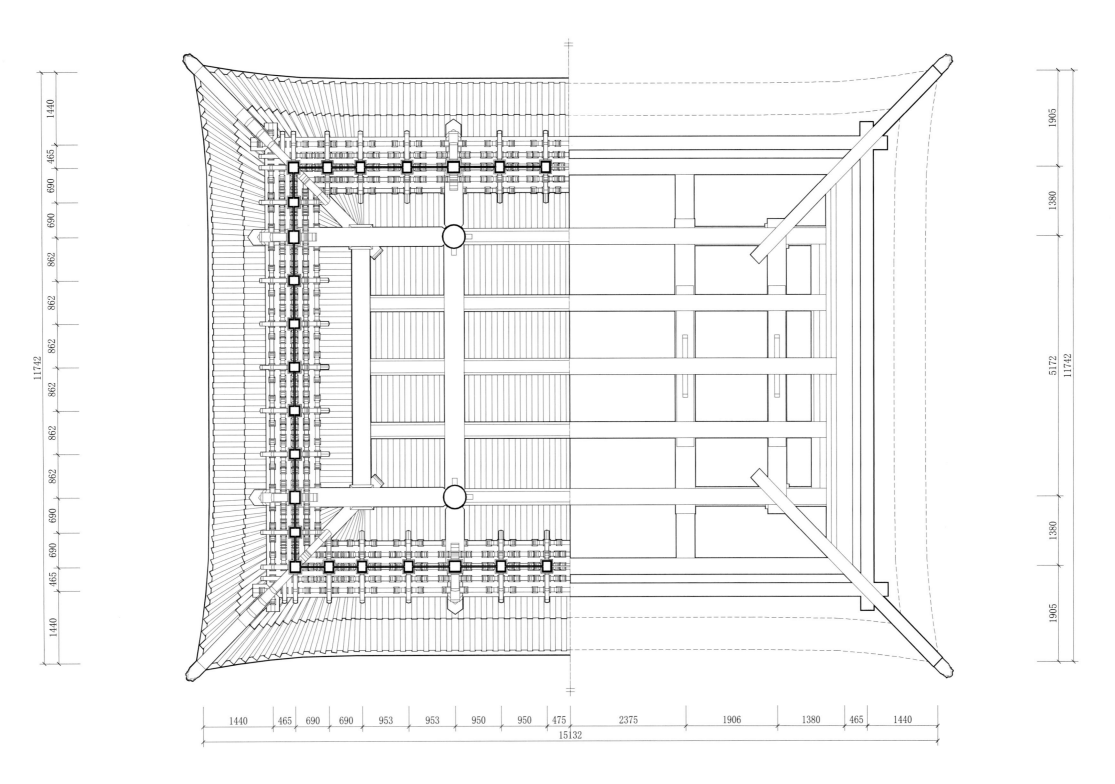

启圣寝殿梁架仰视及俯视图

Plan of framework of Qisheng resting hall as seen from below

殿庭后院

The Backyard

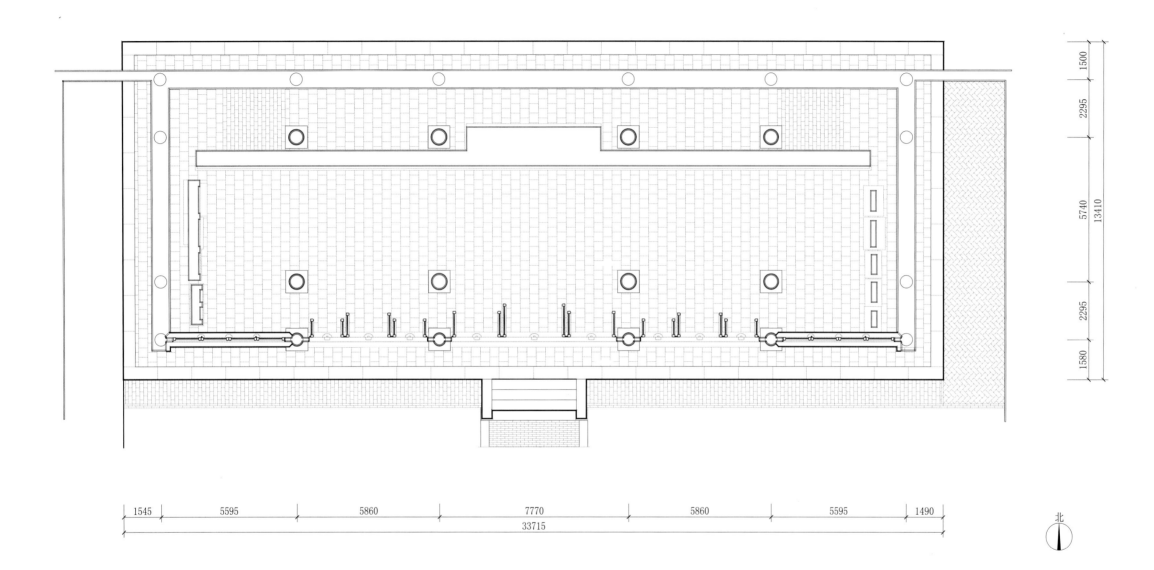

1500
2295
5740
13410
2295
1580

1545　5595　5860　7770　5860　5595　1490
33715

北

圣迹殿平面图
Plan of Shengji Hall

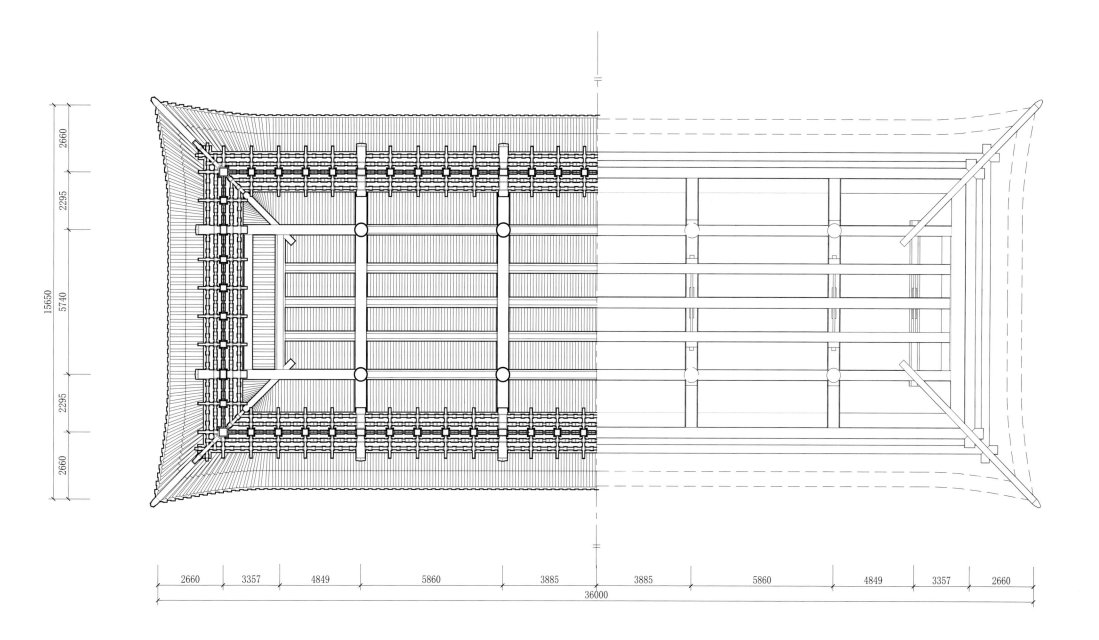

圣迹殿梁架仰视及俯视图
Plan of framework of Shengji Hall as seen from below

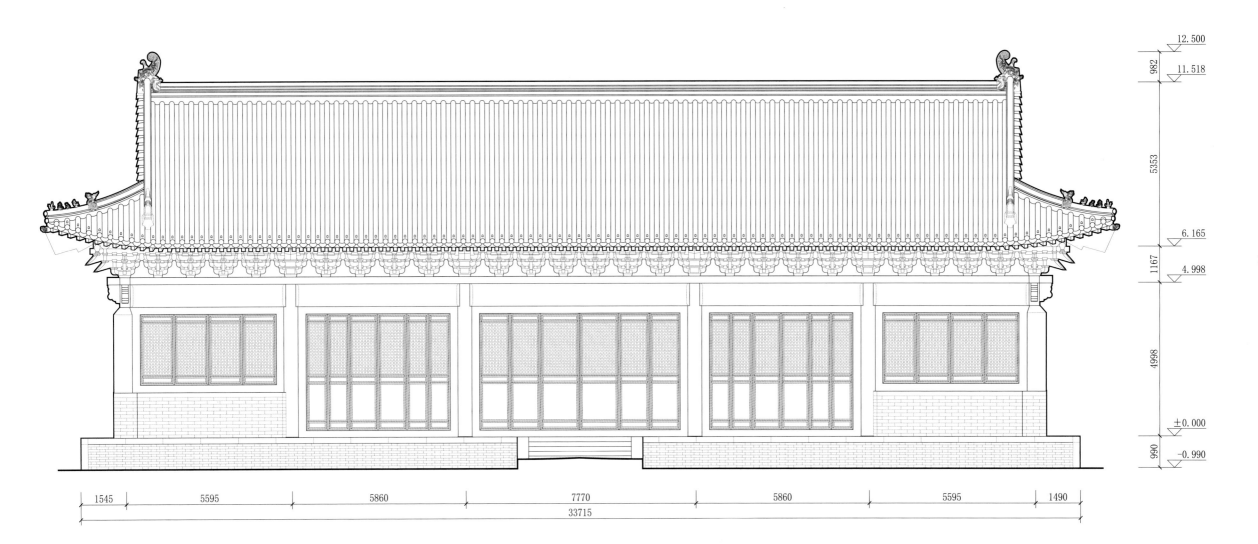

12.500

982   11.518

5353

6.165

236

1167   4.998

4998

±0.000

990   -0.990

1545   5595   5860   7770   5860   5595   1490

33715

圣迹殿正立面图
Front elevation of Shengji Hall

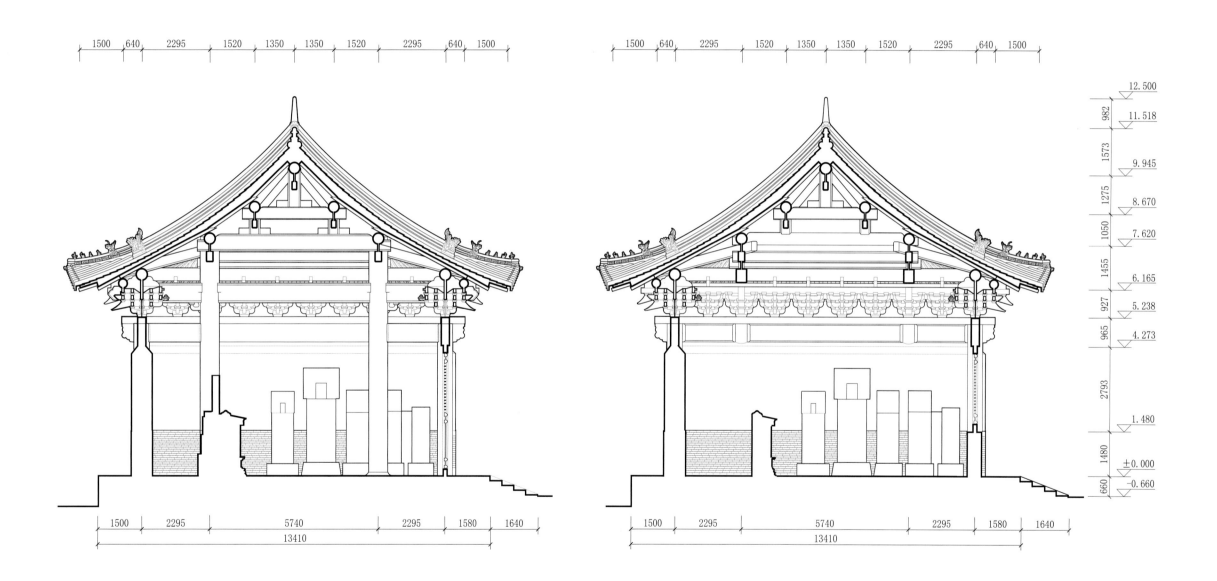

圣迹殿明间剖面图
Section of central bay of Shengji Hall

圣迹殿梢间剖面图
Section of second-to-last bay of Shengji Hall

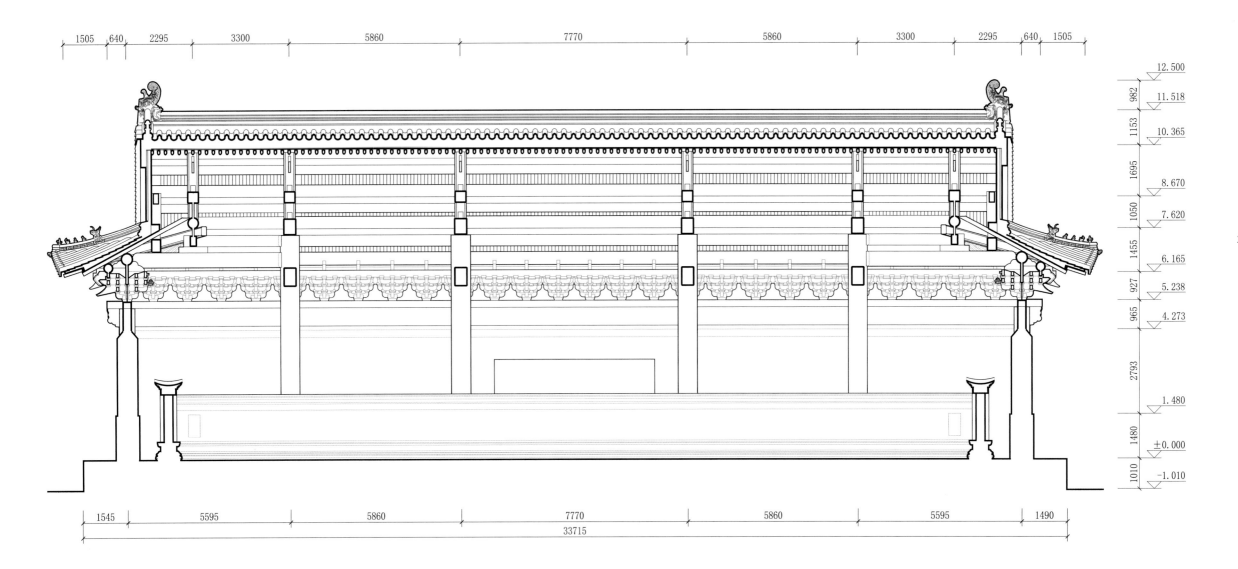

1505 640 2295 3300 5860 7770 5860 3300 2295 640 1505

12.500
982 11.518
1153 10.365
1695 8.670
1050 7.620
1455 6.165
927 5.238
965 4.273
2793 1.480
1480 ±0.000
1010 -1.010

1545 5595 5860 7770 5860 5595 1490
33715

圣迹殿纵剖面图
Longitudinal section of Shengji Hall

圣迹殿院门
The Gate of Courtyard with Shengji Hall

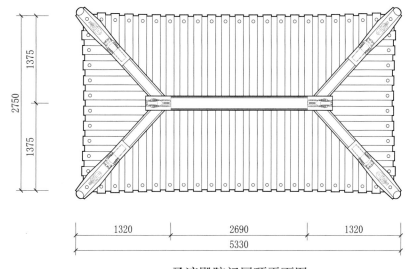

圣迹殿院门屋顶平面图
Plan of roof of the gate of courtyard with Shengji Hall

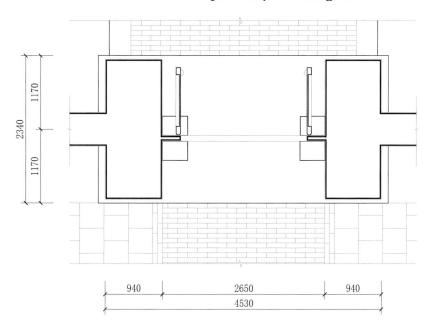

圣迹殿院门平面图
Plan of the gate of courtyard with Shengji Hall

北

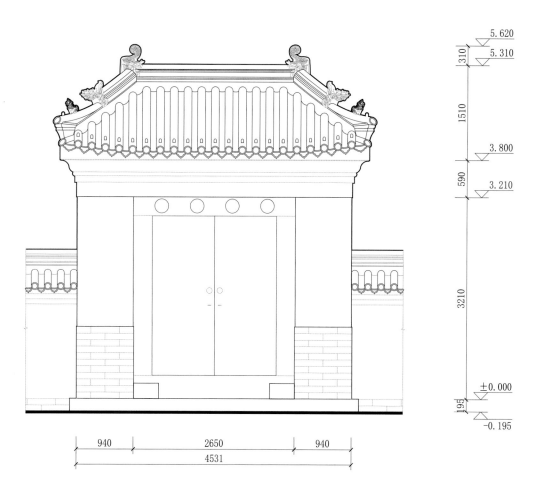

圣迹殿院门正立面图
Front elevation of the gate of courtyard with Shengji Hall

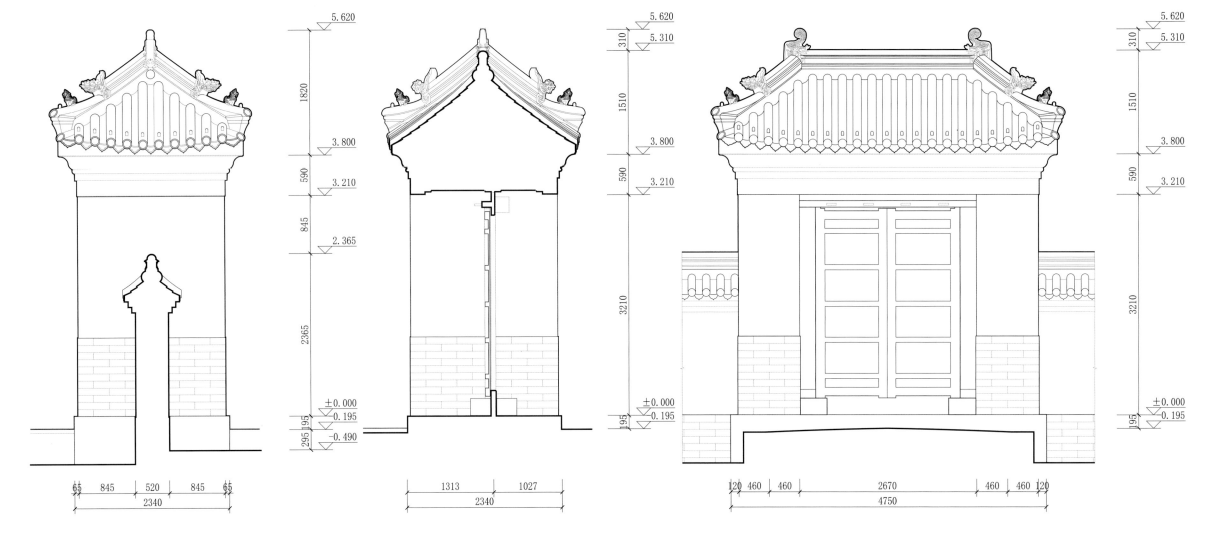

圣迹殿院门侧立面图
Side elevation of the gate of courtyard with Shengji Hall

圣迹殿院门横剖面图
Cross-section of the gate of courtyard with Shengji Hall

圣迹殿院门背立面图
Rear elevation of the gate of courtyard with Shengji Hall

圣迹殿西门
West Gate of Shengji Hall

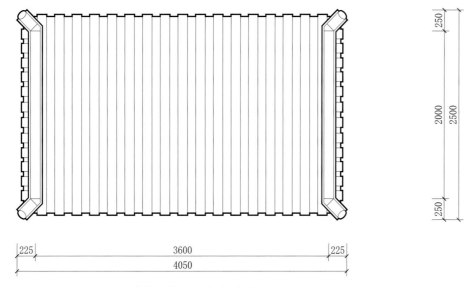

圣迹殿西门屋顶平面图
Plan of roof of west gate of Shengji Hall

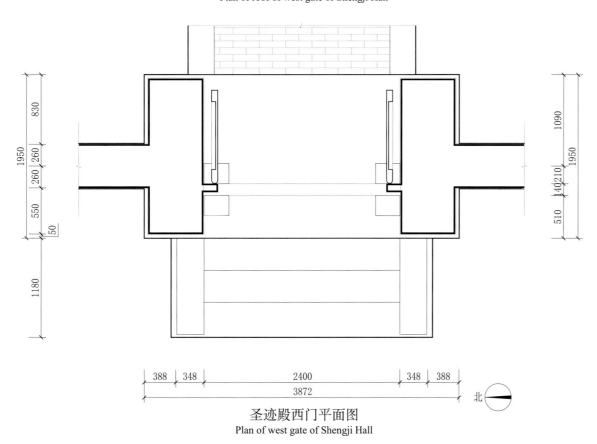

圣迹殿西门平面图
Plan of west gate of Shengji Hall

北

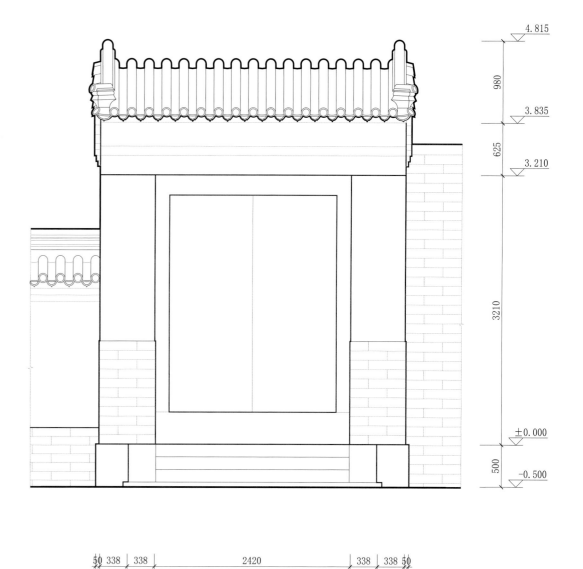

圣迹殿西门正立面图
Front elevation of west gate of Shengji Hall

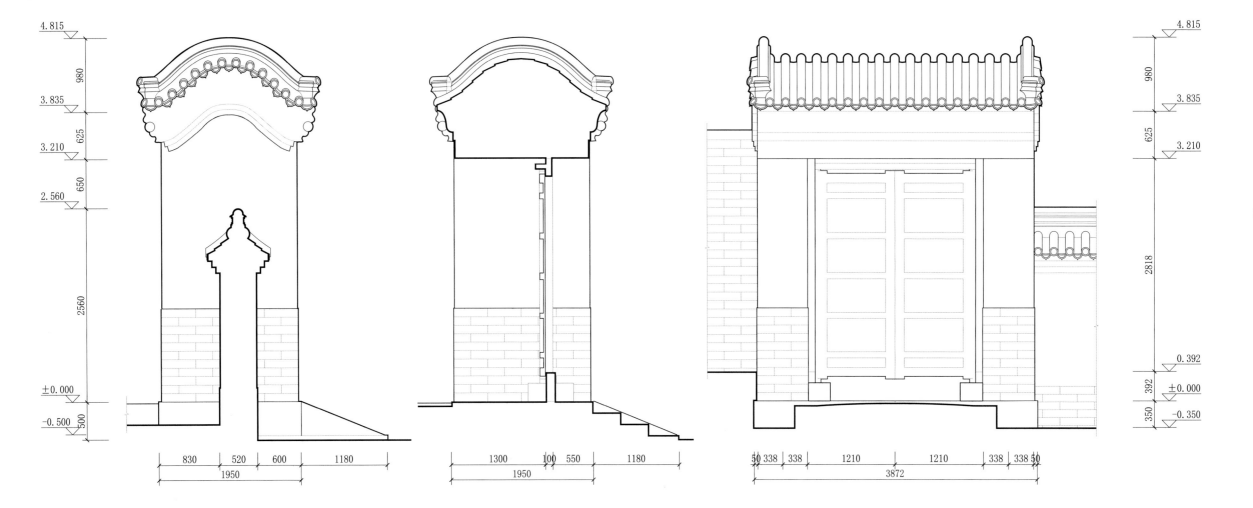

4.815

980

3.835

625

3.210

650

2.560

2560

±0.000

-0.500

500

830　520　600　1180

1950

圣迹殿西门侧立面图
Side elevation of west gate of Shengji Hall

1300　100　550　1180

1950

圣迹殿西门横剖面图
Cross-section of west gate of Shengji Hall

4.815

980

3.835

625

3.210

2818

0.392

392　±0.000

350　-0.350

50 338　338　1210　1210　338　338 50

3872

圣迹殿西门背立面图
Rear elevation of west gate of Shengji Hall

神庖
Sacred abattoir

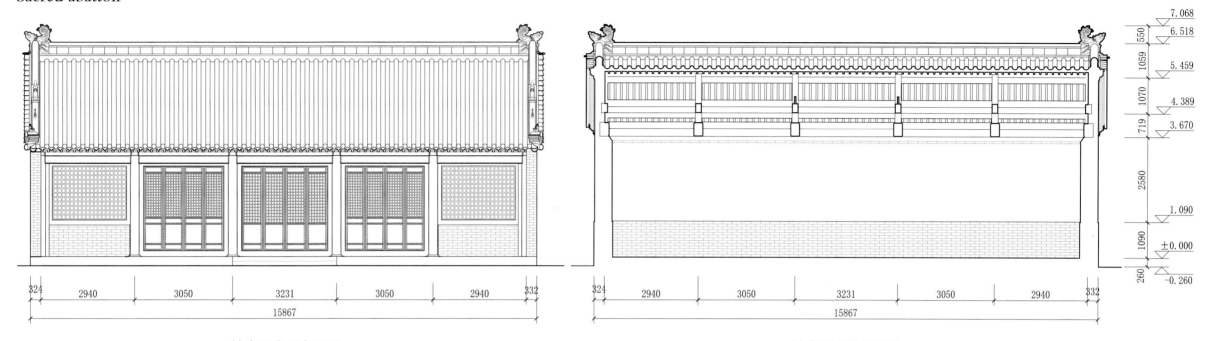

神庖正房正立面图
Front elevation of principal room of Sacred abattoir

神庖正房纵剖面图
Longitudinal section of principal room of Sacred abattoir

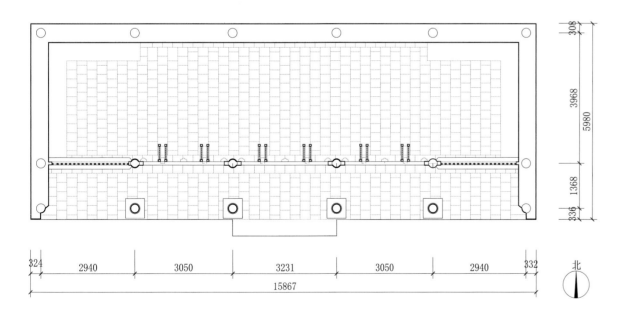

神庖正房平面图
Plan of principal room of Sacred abattoir

北

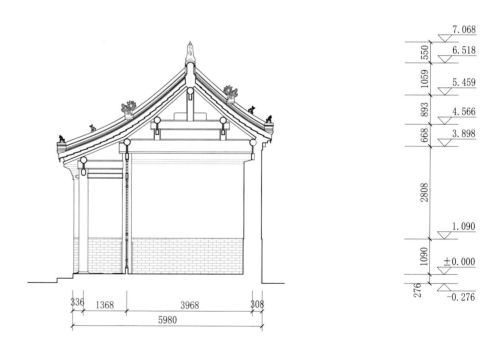

神庖正房横剖面图
Cross-section of principal room of Sacred abattoir

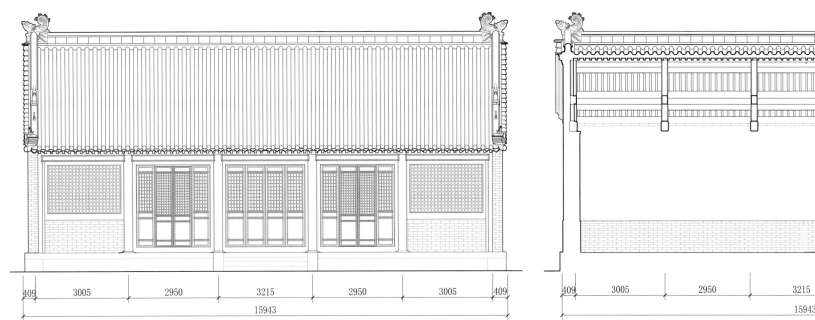

神厨正房正立面图
Front elevation of principal room of Sacred kitchen

神厨正房纵剖面图
Longitudinal section of principal room of Sacred kitchen

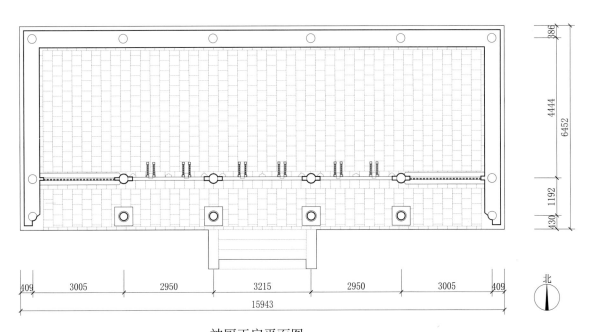

神厨正房平面图
Plan of principal rooms of Sacred kitchen

北

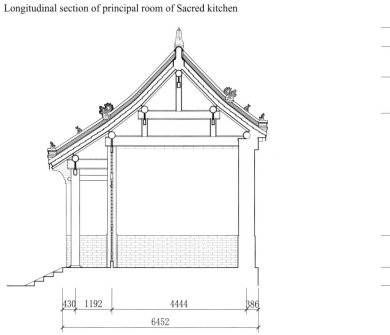

神厨正房横剖面图
Cross-section of principal rooms of Sacred kitchen

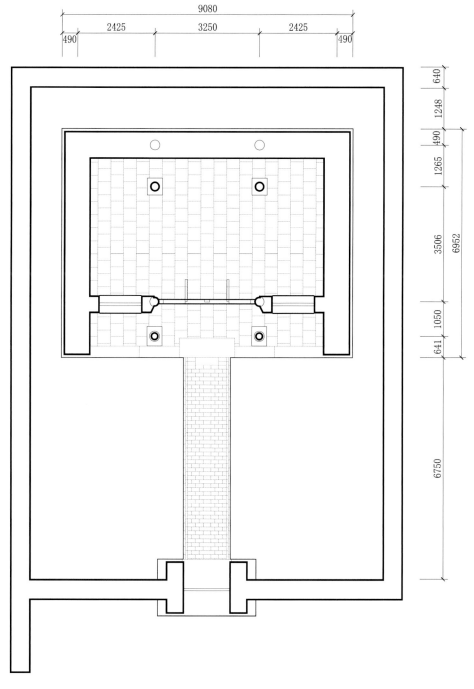

后土祠平面图
Plan of God of the Earth Shrine

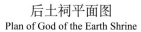

北

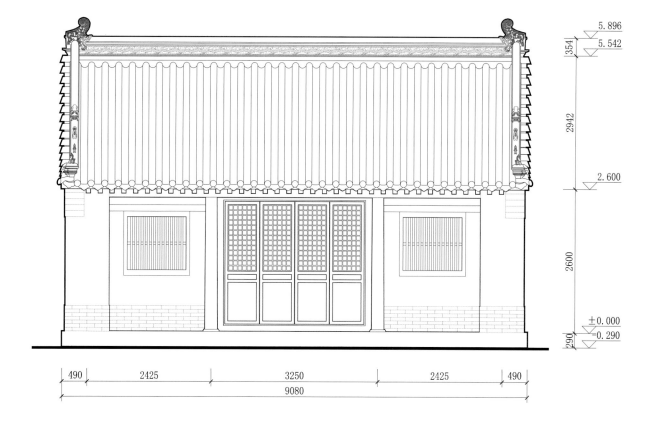

后土祠正立面图
Front elevation of God of the Earth Shrine

参考文献
References

［一］南京工学院建筑系，曲阜文物管理委员会合著．曲阜孔庙建筑［M］．北京：中国建筑工业出版社，1987．

［二］骆承烈汇编．石头上的儒家文献（上、下）［M］．济南：齐鲁书社，1999．

［三］王贵祥．阙里孔庙建筑修建史札［J］．建筑史，2008(00)．

［四］郭满．方志记载折射出的中国古代古迹观念初探［D］．天津：天津大学建筑学院，2013．

［五］王巍，吴葱，韩涛．论遗产记录与真实性［J］．建筑师，2016 (2)．：73-76．

List of Participants Involved in Surveying and Related Works

Name List of Participants from Tianjin University:

Tutors: WANG Qiheng, WANG Wei, WU Cong, BAI Chengjun, ZHUANG Yue, ZHANG Wei

Graduate Students: CAO Peng, WEN Yuqing, YONG Xinqun, WANG Jianghua, LI Xiaodan, DING Yao, WU Dongfan

Undergraduate Students (in alphabet order):

Grade 1996: GE Bin, LIU Xiangfeng, MIAO Bo, TANG Xu, WANG Jing, YAN Kai, ZHU Lei

Grade 1998: CUI Jing, HOU Yanting, LI Jianping, SONG Yuanyuan, WANG Keyao, WU Xiaodong, XIE Dan, ZHANG Long, ZHENG Maoen

Grade 1999: BAI Chen, CHEN Shuanglin, CHEN Yabin, CHEN Yunmiao, DAI Daijun, FANG Wen, FENG Xiaomeng, GONG Xiaolei, GUAN Zhuorui, GUO Aiyuan, GUO Huazhan, GUO Wenhui, HAN Jizheng, HAO Yajie, HE Beijie, HOU Yuefeng, HU Jingying, HU Zhiliang, HUANG Tingdong, JI Chengke, JIN Huiqing, LI Huifeng, LI Hongyu, LI Hui, LI Jirui, Li Rong, LI Xiaofeng, LI Zengquan, LIANG Zhe, LIU Dingwei, LIU Lei, LIU Li, LIU Lisen, LIU Xiaohuan, LIU Xiaoming, LIU Zezhou, LUO Ping, MA Yan, SHANG Hai , SHI Zhigang, SHI Kai, SONG Yongcheng, SU Meng, TAN Xiaoge, TANG Shuo, TU Luoya, WANG Danhui, WANG Jun, WANG Kuan, WANG Nan, WANG Ping, WANG Qirui, WANG Tao, WANG Zhe, WU Haotian, XIE Qi, XIE Zhang, XU Xin, YANG Huan, YANG Jing, YANG Wenyao, YUAN Yi, ZHANG Chunmei, ZHANG Jintao, ZHANG Mengluan, ZHANG Qing, ZHANG Rong, ZHANG Wei, ZHANG Xinnan, ZHANG Yuhui, ZHANG Yuanwang, ZHAO Boyang, ZHAO Qingdong, ZHEN Yi, ZHENG Guodong, ZHENG Xin, ZHOU Jing, ZHOU Zhi, ZHU Xuesong, ZHUANG Yuan

Grade 2000: BAI Wei, HE Jianxiong, ZHAO Guoqiu, ZHONG Lingyuxiu

The Publication of the Drawing Collation Personnel

Drawing Check and Proof: WU Cong, DING Yao

Drawing Arrangement and Modification: WANG Qi, WANG Chu, WANG Yi, YANG Jie, HAN Sai, WANG Wei, FAN Xiaofeng

English Translator: WU Cong

English Proofreader: GUO Han

图书在版编目（CIP）数据

曲阜孔庙 = THE QUFU CONFUCIAN TEMPLE：汉英对照 / 王其亨主编；吴葱，丁垚编著 . — 北京：中国建筑工业出版社，2019.12
（中国古建筑测绘大系 . 祠庙建筑）
ISBN 978-7-112-24552-9

Ⅰ. ①曲… Ⅱ. ①王… ②吴… ③丁… Ⅲ. ①孔庙－建筑艺术－曲阜－图集 Ⅳ. ① TU-882

中国版本图书馆 CIP 数据核字（2019）第 284463 号

丛书策划 / 王莉慧
责任编辑 / 李鸽　刘川
英文校译 / 郭涵
书籍设计 / 付金红
责任校对 / 王烨

中国古建筑测绘大系·祠庙建筑

曲阜孔庙

天津大学建筑学院
曲阜市文物管理委员会　合作编写

王其亨　主编　吴葱　丁垚　编著

Traditional Chinese Architecture Surveying and Mapping Series:
Shrines and Temples Architecture
THE QUFU CONFUCIAN TEMPLE
Compiled by School of Architecture, Tianjin University &
Qufu Municipal Administration of Cultural Heritage
Chief Edited by WANG Qiheng
Edited by WU Cong, DING Yao

*

中国建筑工业出版社出版、发行（北京海淀三里河路 9 号）
各地新华书店、建筑书店经销
北京海视强森文化传媒有限公司制版
北京雅昌艺术印刷有限公司印刷

*

开本：787 毫米 ×1092 毫米　横 1/8　印张：33　字数：825 千字
2022 年 10 月第一版　　2022 年 10 月第一次印刷
定价：**248.00** 元
ISBN 978-7-112-24552-9
（35231）